BUILDER–TESTED | CODE APPROVED

FOUNDATIONS
AND CONCRETE WORK

FROM THE EDITORS OF **Fine Homebuilding**

The Taunton Press

The Taunton Press
Inspiration for hands-on living®

The Taunton Press Inc., 63 South Main St., PO Box 5506, Newtown, CT 06470-5506

e-mail: tp@taunton.com

Editor: Carol Kasper

Copy editor: Seth Reichgott

Indexer: Jay Krieder

Cover design: Alexander Isley, Inc.

Interior design: Cathy Cassidy

Layout: Kimberly Shake

Front cover photographer: Justin Fink, courtesy *Fine Homebuilding*, © The Taunton Press, Inc.

Back cover photographer: Roe A. Osborn, courtesy *Fine Homebuilding*, © The Taunton Press, Inc.

Taunton's For Pros By Pros® and Fine Homebuilding® are trademarks of The Taunton Press Inc., registered in the U.S. Patent and Trademark Office.

Library of Congress Cataloging-Publication-Data

Foundations and concrete work / from the editors of Fine homebuilding.

   p. cm. -- (Taunton's for pros by pros)

  ISBN 978-1-56158-990-6

 1. Concrete footings. 2. Foundations--Materials. 3. Concrete construction.

TH5201.F68 2007

690'.8--dc22

                         2007023515

Printed in the United States of America

10 9 8 7 6 5 4 3 2 1

The following manufacturers/names appearing in *Foundations and Concrete Work* are trademarks: Aberdeen Group®, ABT, Inc. ®, Akzo®, Allan Block®, Amoco®, Amoco Amocor®, Basement Systems®, Belgard®, Bigfoot Systems®, Bituthene®, Bobcat®, Bosch®, Bostik®, Castleblock®, Delta®, DeWalt®, Dow®, Dryvit®, Enkadrain®, Foamular Insul-Drain®, GeoTech®, Grout-Aid by Sika®, Henry® Company, Hitachi®, Hilti®, Hydrocide®, iForm®, Insul-Drain®, Integra Spec®, Karnak®, Keystone®, Kodak®, Kure-N-Seal®, Lego®, Linq®, Makita®, Mameco® Paraseal, Mar-Flex®, Master Builders®, Masterseal®, Metabo®, Milwaukee®, Miraclay®, Miradrain®, Mirafi®, Mosaic®, Noble-seal®, Paraseal®, Pecora®, Polydrain®, Polyguard®, Polyken®, Pisa2®, Rastra®, Revers-A-Cap®, Reward®, Roman Pisa®, Rub-R-Wall®, Safe-coat®, Sashco®, Sawzall®, Set 'N Stone®, Sika® Corp., Sikaflex®, Simpson®, Sonneborn®, Sonolastic®, Sonotube®, Stabila®, Styrofoam®, Sunflex®, Surform®, Sure Klean®, System Platon®, Takeuchi®, Thermadry®, Thoroseal®, ThoroSystem Products®, Tuff-N-Dri®, Tu-Tuf®, Tyvek®, Unilock®, Versa-Lok®, Vulkem®, Warm-N-Dri®, Zoeller®

Special thanks to the authors, editors, art directors,

copy editors, and other staff members of *Fine Homebuilding*

who contributed to the development of the articles in this book.

CONTENTS

PART 3: WATERPROOFING

PART 4: RETROFITTING FOUNDATIONS

PART 5: CONCRETE IN THE LANDSCAPE

INTRODUCTION

After graduating from college I did what anybody with a B.A. in English would do: I went to work as a form carpenter on a concrete crew. I spent my days hauling heavy, smelly, cement-encrusted, oil-soaked forms in and out of trenches. You set up the forms and take them down. You scrape them off. You do it again. And usually you're walking around in mud-caked boots that feel like you're wearing a pumpkin on each foot.

I worked for a boss who delighted in ordering concrete before we were ready. He'd start filling the forms while I was still working frantically to nail them together. I still wince when I see a ready-mix truck coming down the road, but my year on that crew taught me a lot about pouring concrete and building foundations. I learned how to place anchor bolts so they could withstand a hurricane. I learned how to use rebar and wire mesh to strengthen a slab, and the consequences of not bracing forms properly (you don't want to know).

Fortunately, you don't have to spend a year walking around in muddy boots to learn what I did. This book contains all of that knowledge, and then some. Collected here are 28 articles from past issues of *Fine Homebuilding*. Written by builders from all over the country, these articles will help ensure that you can build a strong foundation that's dry, warm, square, and level. Do yourself one favor, though: Wait until the forms are done before you order the concrete.

Kevin Ireton, editor
Fine Homebuilding

Avoiding Common Masonry Mistakes

■ BY THOR MATTESON

At the engineering firm where I work, the past few years have brought us about a dozen jobs retrofitting designs for relatively new buildings that were structurally deficient or failing for one reason or another. Typical was the work we did on a poorly designed office building. Improperly placed rebar substantially reduced the strength of a critical grade beam. After a good deal of excavation, we epoxied dowels into the old grade beam and reinforced it with 10 yd. of new concrete.

Now I notice structural problems everywhere I look. I can't even go to the supermarket without wincing ever so slightly at the shrinkage cracks in its concrete-block walls (see the photo on the facing page). Although these problems are how I make a living, many of them could have been avoided.

Concrete Strength Relies on the Right Mix and Reinforcement

Reinforced concrete is barely a hundred years old, and engineers are still refining their assumptions of its properties. Yet some contractors (even large governmental agencies) have not changed their methods in the past 30 to 40 years.

Concrete alone lacks appreciable tensile strength. Steel reinforcing, or rebar, used in concrete cannot withstand compressive force by itself. Combining the strengths of concrete and steel produces the required structural properties (see the sidebar on p. 14).

This wall made its own contraction joint. Quarter-inch contraction joints made of a high-grade elastomeric sealant allow a block wall to shrink as water dissipates from the grout, instead of cracking. This wall should have had these control joints every 15 ft. to 20 ft.

Rebar Must Be Detailed Carefully at Corners

At intersections and corners of concrete footings and walls, problems can arise if reinforcement is improperly placed. Steel bars must overlap the correct length and should hook around perpendicular reinforcing. At corners, inside bars should cross, run past each other, and extend toward the far side of the footing. Otherwise, the inside bar in the corner lacks sufficient embedment, and the concrete then may pop out. The three pairs of drawings below are plan views of footings.

Right

CORNER

Rebar

Vertical bars

Minimum 12-in. hook

Wrong

Shaded area can pop out.

Right

INTERSECTION

Wrong

Right

OBTUSE CORNER

Wrong

Section through stepped footing

Right

2 ft. minimum

1 ft. maximum

Wrong

Anyone who works with concrete knows that steel reinforcement provides tensile strength. But even experienced builders and designers commonly overlook an obvious consequence of this fact. Rebar must extend into concrete deep enough to develop that tensile strength. Instead of pulling out of the concrete, the bars will start to stretch.

Problems commonly arise at points where rebar changes direction and at intersections. Take, for example, a corner in a footing that has two horizontal rebars. Workers often place the outer bar wrapping around the outside of the corner, which is correct. But then they wrap the inside bar around the inside of the corner, losing a few essential inches of development length. Inside bars at corners should cross, run past each other, and extend toward the far side of the footing (see the drawings at left).

Rebar Has to Go Deep Enough to Do the Job

Straight reinforcing bars can develop sufficient bond if they extend far enough into the concrete. When you don't have thick enough concrete (such as at a wall corner or at a T-intersection), a hook at the bar's end may substitute for the lack of available embedment. If a perpendicular bar is placed inside the hook, it spreads the force to a greater area of concrete. (Usually there is a whole row of hooked bars, and a single bar can run through all of the hooks.) We like to see a bar inside the bend at any change in bar direction.

Sometimes the tail of a hook may not fit where it's shown on the plans. Usually you can rotate the tail of the hook to clear obstacles. As long as the hook extends into concrete to develop sufficient anchorage, it's doing its job. You should check with the designer first, though. Sometimes hook tails need to lap with other bars.

Less Water Means Less Shrinkage and More Strength

Nuisance cracking in concrete has two major causes: shrinkage during the curing process and thermal expansion or contraction due to temperature swings. Generally, shrinkage is caused as the water in the concrete gradually dissipates. As much as one-third of the water, or 5 percent of the total volume of the concrete, can dissipate.

The strongest concrete uses only enough water to hydrate all of the cement in the mix. Excess water leaves space in the concrete when it evaporates, making the concrete less dense and therefore weaker. (Lost water affects the concrete's structure on a molecular level, unlike the tiny bubbles that are left in air-entrained concrete.) But strong concrete is worthless if you can't place it, and getting good workability requires more water.

Good workability means that you can consolidate the concrete around reinforcement and into corners by vibration. It does not mean that the concrete flows there by itself. Adding a water-reducing agent to the concrete can make it easier to work without using more water.

Reducing the amount of water in the mix also makes the cured concrete less permeable to air, water, and salts. Although air seems harmless, the carbon dioxide it carries can react with water in a process called carbonation. In carbonation, water and carbon dioxide combine to form carbonic acid, a weak acid. Over time, enough acid lowers the concrete's alkalinity to the point where steel corrodes much more readily. Using a drier mix and consolidating it well can forestall carbonation for a long time.

Substituting Pea Gravel for Sand Can Prevent Problems

Using a high proportion of sand makes concrete easier to finish, but it causes problems, including increased shrinkage and reduced strength. A mix with as little sand as possible helps ease these problems.

Reducing the amount of sand makes concrete harder to work, but it also means fewer shrinkage cracks. Here's why: To make concrete workable, water must coat the surfaces of all of the aggregate. For a given weight of aggregate, large pieces will have considerably less surface area than small grains, and thus require less water to provide lubrication. Substituting pea gravel for some of the sand in the mix allows you to reduce the amount of water needed for workability. Less water is left free to evaporate, and the concrete will crack less.

When you order a load of concrete, you can specify less sand and more gravel. Your local batch plant should have various mix designs on file so that you can specify a 60–40 or 65–35 mix (the ratio of gravel to sand).

Less water also reduces potential reactive aggregate problems. Some aggregates will gradually react with the alkaline cement and expand slightly in the process. When this happens, the concrete's own ingredients break it apart, and it gradually disintegrates. Although coarse reactive aggregates can be sorted out economically, a similar process does not exist for sand, so the less sand the better.

Reducing the amount of sand makes concrete harder to work, but it also means fewer shrinkage cracks.

Special Considerations for Slabs on Grade

Concrete slabs need to have a firm, even, well-compacted substrate, which begins with properly prepared original ground. Remove all sod, stumps, roots, other organic matter, large rocks, and wet, mushy soil. The soil must be compacted thoroughly and evenly. If you have unstable soil, plan on hiring an engineer or ending up with a cracked, buckled slab.

Moisture intrusion through slabs varies greatly in different regions of the country. In most cases, the American Concrete Institue recommends not using vapor barriers. This subject creates great debate between all involved.

Current recommendations suggest using "crusher-run" rock (also called "road base") underneath slabs on grade. Sand does not perform as well as road base. If local conditions mandate the use of a vapor barrier, it should be placed beneath the layer of crusher-run rock.

Rebar Performs Better Than Wire Mesh in Slabs

Steel rebar and wire mesh both serve the same function: They add structural reinforcement to the concrete. Rebar is only slightly more expensive than mesh but easier to keep centered in the slab. If mesh

Wire mesh can end up at the bottom of the slab. This core sample of a wire-mesh reinforced concrete slab illustrates what happens when concrete is poured over the mesh. Rather than staying in the middle of the slab, where it strengthens the slab, the mesh gets pushed to the bottom, where it does little good.

Rebar on Chairs Stays in Center of Slab

In concrete slabs, steel-bar reinforcement works better than wire-mesh reinforcement because it stays where it's put. It's also important to have a firm, even bed beneath the slab and to cover the substrate with a plastic vapor barrier.

#3 rebar at 18 in. o. c.

Chairs to support rebar

Minimum 2 in. between steel and ground

Bed of sand, gravel, or crushed stone beneath poly

Control Joints Help Control Cracking

Shrinkage cracks appear in any slab. But you can keep them small and govern where they appear by building in control joints. A control joint works like perforations in a piece of paper. The joint is a line of weakness in the slab that eventually becomes a crack.

If you place the joints at 12-ft. to 15-ft. intervals, most cracking occurs along the joints. For large industrial or commercial slabs, control joints usually are sawn into the concrete. On smaller jobs it may not be worthwhile to bring a concrete saw on site. In such cases, long pieces of plastic extruded in T-shaped cross section can form the joints. Concrete finishers force the stem of the T into the slab. To ensure that cracks occur at the control joints, their depth should be at least one-quarter of the slab's thickness.

is properly placed, it should do just as good a job of strengthening the concrete as rebar.

Reinforcement should stay centered in the slab and not trip concrete workers as they move about. However, if the reinforcement is wire mesh, there's no way to avoid trampling it while placing the concrete. So the mesh gets embedded in the sand beneath the slab. Although it's not impossible to keep the mesh centered uniformly in the slab, it is difficult.

Placing a grid of #3 rebar at 18-in. centers provides adequate slab reinforcement and places for nimble feet to step. Supporting the bars on metal "high chairs" or precast concrete cubes (called "dobies" where I live) keeps them in the proper position as concrete is poured over them. The cost difference between mesh and rebar is almost negligible, and the reinforcement ends up where it belongs (see the drawing and photo on the facing page).

Reinforcement falls into two main categories: structural and shrinkage/temperature. Structural reinforcement provides strength to resist bending, compression, or tensile loads. Shrinkage/temperature reinforcement reduces the concrete's tendency to crack as it dries or as it contracts or expands due to temperature changes.

For the latter, chopped fibers of polypropylene, nylon, steel, glass, or the like may be added to the concrete mixture. Because the fibers are automatically distributed throughout the concrete during mixing, there are no concerns about proper placement. For most slabs on grade that require no structural reinforcement, fiber reinforcing can be placed in the mix at the batch plant.

Tips on Placing and Working Concrete

Concrete-form construction is beyond the scope of this article. But when you build forms, build them stronger than necessary. Brace them well; expect to climb all over them while carrying an ornery vibrator.

Secure anchor bolts and other inserts to the forms in their proper locations before the pour. (Poking bolts into wet concrete disturbs the aggregate and gives a weaker bond than pouring the mix around the bolt.) If you use a form-release agent, apply it to the form boards only, not to the reinforcement.

The Right Way and the Wrong Way to Place Concrete

If concrete is dropped from a height greater than 4 ft. into a form or permitted to fall freely over reinforcement, the aggregate can separate from the concrete or honeycomb at the bottom. When filling forms from a chute, use a hopper to deliver concrete to the bottom of the forms. When using a pump, feed the hose to the bottom of the form.

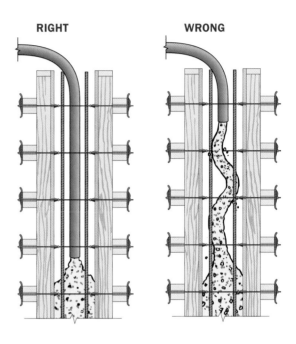

Always place concrete as close to its final position as possible. When concrete comes out of the chute or the pump hose, it should not free-fall more than 4 ft., or clatter off the rebar or forms (see the drawings above). Either of these conditions can cause the coarse aggregate to separate from the rest of the mix, resulting in concrete that's not uniform. Preventing this condition in tall walls or columns usually requires that a concrete pumper dispense the concrete from a hose, which can maneuver close to the bottom of the forms.

Building codes require consolidating all structural concrete to eliminate voids. The most effective way to do this is with a concrete vibrator. Turn the vibrator on and insert it into the concrete as quickly as possible. To vibrate the first lift poured, insert the vibrator all the way to the bottom. Then withdraw it slowly, about 3 in. per second. The goal is to allow all of the trapped air

bubbles to get out of the concrete. The bubbles move up slowly; if the vibrator head moves faster than they do, they will remain trapped. Vibrate any additional lifts the same way, but extend the end of the vibrator about 6 in. into the previous lift.

The vibrator influences a circular area of concrete, whose size depends on the power of the tool. These circles of influence should overlap. Use a regular pattern and consolidate the concrete. Do not insert the vibrator at haphazard angles or use it to move concrete in the forms.

Keep the Concrete Wet and Warm

Builders always want to hurry to strip the forms, but leaving them in place a few days holds the moisture in the concrete. A week of wet curing would make any engineer happy. Keep the concrete wet by covering it with plastic or wet burlap, or with spray from a fog nozzle. Slabs also can be flooded. Curing compounds that seal in moisture when sprayed on concrete are available.

Rapid drying or freezing severely reduces concrete's strength and results in weak slab surfaces. A little time invested in proper curing protects the finished product you worked hard for.

For further information, the American Concrete Institute publishes several references: ACI 302.1, "Guide for Concrete Floor Slab Construction," should be in every contractor's library. The Aberdeen Group® offers several publications and references intended for contractors and builders. "Design and Control of Concrete Mixtures" by the Portland Cement Association offers 200 pages of information, including mixing, placing, finishing, and testing.

Understanding the Principles of Concrete-Block Construction

Reinforced concrete-block construction can produce strong, durable walls efficiently. In this type of construction, three components form the structural system: the blocks and the mortar that holds them together, the reinforcement, and the grout, which is used to fill in the cores in the concrete block.

Grout is a mix of fine gravel, sand, cement, and water. In some areas (not California), mortar is used instead of grout in block cores. However, grout contains coarser aggregate than mortar, which makes it stronger. Building codes require each cell that contains reinforcement to be filled with grout. In seismic zones 3 and 4 (almost all of California and the West Coast), this rule means filling at least every sixth vertical cell (every 4 ft.) and grouting a bond beam at the bottom, middle, and top of an 8-ft. wall. (Bond-beam blocks are made with space for horizontal rebar to lay in them; a course of

these blocks, when reinforced with steel and filled with grout, forms a bond beam.) Additional bars at openings, or where required by the designer, often decrease the spacing to 32 in. or even 24 in. Grouting all of the cells is usually easier than trying to block off the cells where grout is not required, so most walls we see are solid-grouted. In essence, concrete blocks or withes of brick just serve as the forms for a fine-aggregate concrete (grout) wall. Your wall should solidly attach to the footing and act as a monolithic unit, not a stack of separate blocks. Also, cores should be grouted after the wall is built, not as you go, which would mean cold joints.

Clean-Outs Ensure That the Wall Bonds to the Footing

Openings at the base of cells that will receive grout allow you to remove construction debris before pouring (see the drawing on p. 12). When all of the blocks are laid, you can remove the mortar droppings, nails, tape measures, cellular phones, and the like that have fallen to the bottom of each cell. Initially pouring an inch or so of sand at the base of the clean-out prevents fresh mortar from sticking to the footing. These chunks decrease the bond between the grout and the footing. Use a length of rebar to chip off any mortar tags that project more than ½ in. into the cells. After removing all the debris, seal the clean-outs. Let the mortar set a few days (to prevent blowouts) before the grout is poured. Building codes require clean-outs in cells containing reinforcement if a grout pour will be more than 5 ft. high. For shorter lifts, you can suck debris out with a shop vacuum, although you may have trouble winding past all of those bars.

Rapid drying or freezing severely reduces concrete's strength and results in weak slab surfaces.

Ample Reinforcement Leaves Room for Vibrating and Has Built-In Joints

A single horizontal bar centered in this composite drawing of various walls would have made it difficult to vibrate the grout. Two bars spaced apart provide greater strength and leave room for the vibrator head to snake down to the bottom. Also, a ¼-in. contraction joint made of special caulking allows the wall to shrink without cracking. The horizontal reinforcement that runs through the course of bond-beam units should break at the joint.

Two bars of horizontal reinforcement

8-in. by 8-in. by 16-in. bond-beam block

One bar of vertical reinforcement

Cell to be grouted

Wire cradle to hold up horizontal reinforcement and separate vertical bars

¼-in. contraction joint made of high-grade elastomeric sealant, every 15 ft. to 20 ft.

Horizontal ladder-type joint reinforcement mortared into wall

8-in. by 8-in. by 16-in. standard block

Clean-outs

CLEAN-OUTS AT BOTTOMS OF CELLS ENSURE GOOD GROUT BOND

Clean-out blocks allow mortar droppings and debris to be removed before grout is poured, letting the grout adhere to the concrete footing. Before the pour, the face of the clean-out block is mortared back in place.

Standard Types of Concrete Block

Here are a few of the more standard types of concrete block. There are many other blocks made for a variety of different functions and in a variety of different sizes, configurations, and textures.

8-in. by 8-in. by 16-in. open end

Normal, or standard, block

8-in. by 8-in. by 16-in. open-end bond beam

8-in. by 8-in. by 16-in. double open-end bond beam

8-in. by 8-in. by 16-in. mortar-less head joint or "speed block"

Ladder-Type Joint Reinforcement Is Better Than Truss-Type

Joint reinforcement can be used instead of bond-beam units that contain horizontal bars. This type of reinforcement consists of two horizontal, parallel wire rods connected by cross ties (see the drawing on facing page). The reinforcement is placed between courses of block or brick, and the side rods get embedded in the mortar. The cross ties may go straight across or zigzag; these cross ties are called ladder and truss types, respectively.

The cross ties of ladder reinforcement should align vertically with the webs of the blocks. Truss-type wire reinforcement is even more difficult to align. We don't recommend using this reinforcement in concrete-block walls because the diagonal reinforcement can block the open cell and make it difficult to insert the vibrator.

Use Pairs of Horizontal Rebar to Make Room for a Vibrator

Rather than using single horizontal bars in bond beams, which lie in the middle of the wall cavity, use pairs of horizontal bars (either smaller bars or the same size at greater spacing). This process leaves more room for the vibrator head. Designers should consider this idea, but if yours hasn't, ask about it before proceeding. In straight runs of wall, open-end units (also known as "A" blocks) allow grout to flow more easily through the block cavities. Using standard units forms air gaps between blocks at head joints.

Admixture Increases Grout Bond to Block

Certain compounds react with grout ingredients to produce a gas. The gas expands, forcing the grout into the porous surface of the block cells. Several brands of admixture are available; most use powdered aluminum as the active ingredient (Grout-Aid by Sika® Corp. or MB612 by Master Builders® Inc.).

Most masons add the admixture once the grout arrives at the site because its working time is limited to about an hour. Although the powder may be added straight into the truck, we require it first to be thoroughly mixed with water. If clumps of powder get pumped into the wall, they can generate enough gas pressure to pop the block apart.

Vibrate Grout Twice

For grout to flow into all of the crevices in a wall, it needs to be as fluid as possible. This wet grout is poured into dry masonry and vibrated immediately. The masonry soaks up excess water. As this occurs, the grout loses volume. The grout may actually shrink away from one side of the grout space. Building codes require reconsolidating the grout with a second vibrating. Vibrating a second time settles the grout fully into the cavities.

I have seen the level of a 4-ft. grout lift (the maximum height allowed) drop by 2 in. during revibration. It's important to wait at least 20 minutes before revibrating so that excess water has time to soak into the masonry. But do not wait so long that the expanding admixture reacts completely or that the grout begins to set up. In mild weather, revibrate within 45 minutes to an hour of placing the grout.

For grout to flow into all of the crevices in a wall, it needs to be as fluid as possible.

More about Concrete Reinforcement

A common gauge of steel's strength is its yield stress. When you apply stress below the yield stress and then release it, the steel returns to its original shape. But when you apply stress greater than the yield stress, the steel begins to deform permanently. The yield stress of reinforcing steel is indicated by its grade. Grade 40 means the yield stress is 40,000 psi. Most reinforcing bars are either grade 40 or grade 60.

U.S. steel mills produce reinforcing bars in 11 sizes, which are denoted by numbers (see the chart below). The numbers in the left-hand column represent the bar's nominal diameter in ⅛-in. increments. So a #4 bar is ½ in. dia. and a #18 bar is 2¼ in. dia. (Since the original publication of this article, rebar is now marked with bar sizes in millimeters.)

Multiplying the bar's yield stress by its cross-sectional area gives its strength. For example, a #4 bar of grade-40 steel will withstand 8,000 lb. of force (0.2 sq. in. x 40,000 psi).

A bar's size and grade are stamped along its length, appearing in a column of symbols, letters, and numbers. The first number to appear is the bar size. For some reason, the steel grade is more deeply hidden. Grade-40 bars have no special or additional marks. Bars of grade-60 steel have the number 60 stamped as the last symbol in the column, or they have an additional rib in their deformed pattern.

For the strength of a bar to be realized, it has to be held in a tight grip by the concrete. This bond strength comes from the bond between the concrete and the rebar, and it depends mostly on the concrete's strength and the holding ability of the rebar. The combination of concrete strength and holding ability of rebar is known as development length. For a #4 grade-40 rebar, the correct development length is the distance the rebar must be

STANDARD REINFORCING-BAR SIZES

To determine the strength of a given size of rebar, multiply its yield stress (40,000 psi and 60,000 psi for grades 40 and 60) by its cross-sectional area.

Bar size		Weight per ft.		Diameter		Cross-sectional area	
	mm	lb.	kg	in.	mm	sq. in.	sq. cm
#3	10	0.376	0.171	0.375	10	0.11	0.71
#4	13	0.668	0.303	0.500	13	0.20	1.29
#5	16	1.043	0.473	0.625	16	0.31	2.00
#6	19	1.502	0.681	0.750	19	0.44	2.84
#7	22	2.044	0.927	0.875	22	0.60	3.87
#8	25	2.670	1.211	1.000	25	0.79	5.10
#9	29	3.400	1.542	1.128	29	1.00	6.45
#10	32	4.303	1.952	1.270	32	1.27	8.19
#11	36	5.313	2.410	1.410	36	1.56	10.10
#14	43	7.650	3.470	1.693	43	2.25	14.50
#18	57	13.600	6.169	2.257	57	4.00	25.80

embedded into the concrete so that when you pull on it with 8,000 lb. of force, the rebar stretches rather than pulls out of the concrete.

The American Concrete Institute's (ACI) model code gives formulas for determining development lengths. For "small" bars (#6 and smaller) and typical residential concrete, the development length is approximately 32 to 36 times the diameter of the bar.

For more information on rebar, contact the Concrete Reinforcing Steel Institute.

Good information right on the bar. The M signifies the mill that made the steel; the 6 means it's #6 rebar; the S means it's made of new billet steel (the type normally found in rebar); the 6 and 0 mean its yield stress is 60,000 psi.

Vertical Contraction Joints Reduce Cracks in Long Walls

Masonry walls shrink in length and height as excess water in them dissipates. It may seem as if you're weakening the wall, but if you don't provide contraction joints to accommodate this shrinkage, the wall makes its own—in the form of cracks. To avoid shrinkage cracks, build long walls in segments no more than 16 ft. long.

Build each segment as if it were an individual wall. Separate the wall segments about ¼ in. Instead of a mortar joint between them, fill the gap with a high-grade elastomeric sealant, such as Sonolastic® from Sonneborn®; or Sikaflex® from Sika Corp. Joints can break up a wall's appearance, but they look better than cracks.

Using some or all of these tips will make your work easier and stronger. For further information, the National Concrete Masonry Association publishes its TEK briefs on a wealth of masonry topics.

Thor Matteson is a structural engineer in Mariposa, California, and is the author of Wood-Framed Shear Wall Construction—An Illustrated Guide. *His website is www.shearwalls.com*

Sources

American Concrete Institute
ACI International
P.O. Box 9094
Farmington Hills, MI 48331
248-848-3700
www.aci-int.org

The Aberdeen Group
426 S. Westgate
Addison, IL 60101
800-323-3550
www.worldofconcrete.com

Concrete Reinforcing Steel Institute
933 N. Plum Grove Rd.
Schaumburg, IL 60173
847-517-1200
www.crsi.org

Master Builders Inc.
23700 Chagrin Blvd.
Cleveland, OH 44122
800-628-9990
www.masterbuilders.com

National Concrete Masonry Association
13750 Sunrise Valley Dr.
Herndon, VA 20171
703-713-1900
www.ncma.org

Portland Cement Association
5420 Old Orchard Rd.
Skokie, IL 60077
847-966-6200
www.cement.org

Sika Corporation
201 Polito Ave.
Lyndhurst, NJ 07071
800-433-9517
www.sika.com

Sonneborn
Chem-Rex Inc.
889 Valley Park Dr.
Shakopee, MN 55379
800-433-9517
www.buildingsystems.basf.com

Working with Rebar

■ BY HOWARD STEIN

Why should you have a spare tire in the bed of your truck? It's not required by law, and the truck will run just fine without it. Using essentially the same logic, some people might ask why we use rebar in our residential foundations. Many houses are built without it (East Coast building codes don't require it), and if a house is built with rebar, building inspectors often don't inspect its placement.

The simple answer is that rebar is cheap insurance against the potential problems that can develop after concrete is poured or, worse, after the foundation has been backfilled. A foundation that has gone wrong is extremely expensive to repair. Just to be safe, my crew and I reinforce concrete in footings and walls, in piers and columns, and in structural slabs, and we also use rebar to tie new work to old.

Concrete used in residential construction is usually specified in a range of compressive strength in 500-lb. increments from 2,500 psi to 4,000 psi. It's obvious that concrete can support phenomenal compression loading. However, when it is under tension or shear forces, concrete has lower values compared with other common construction materials. If the underlying soils are of uneven densi-

ties, differential settling beneath the foundation can cause large cracks in the walls. Concrete is also subject to shrinkage cracks, especially when poured with a high water-to-cement ratio. When properly sized and embedded in concrete, rebar partially compensates for these deficiencies. (Remember that excess water produces lower-psi concrete that is weaker and more prone to shrinkage. Even reinforced with rebar, high-water-ratio concrete shouldn't be used.)

Sizes and Grades of Rebar

Rebar comes in many sizes and grades (see the photo on p. 18). In residential work, we mostly use bar sizes #3, #4, or #5. These sizes translate to the diameter of the stock, measured in ⅛-in. increments; #3 bars are ⅜ in. in dia., #4 bars are ½ in. (½ in.), and #5 bars are ⅝ in. The grades 40 and 60 refer to the yield (tensile) strength (40,000 psi

Cutting and bending rebar is a matter of leverage and muscle. The cutter/bender's head provides a lever and fulcrum that makes simple bends; more complex bends can be made from individual pieces lapped together.

Know Your Rebar

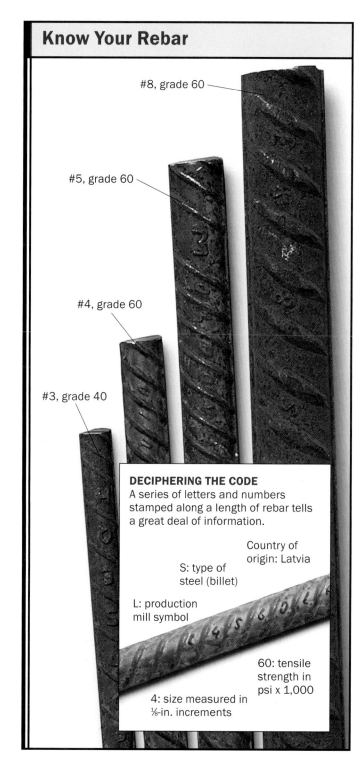

#8, grade 60

#5, grade 60

#4, grade 60

#3, grade 40

DECIPHERING THE CODE
A series of letters and numbers stamped along a length of rebar tells a great deal of information.

Country of origin: Latvia

S: type of steel (billet)

L: production mill symbol

60: tensile strength in psi x 1,000

4: size measured in ⅛-in. increments

and 60,000 psi, respectively). Grade 60 is harder to cut and bend. Both grades are priced the same. The designer usually specifies which one to use for a particular purpose. If the grade is not specified, I buy the softer grade 40 for short lengths and bends, and grade 60 for long straight runs with few or no bends.

Make Sure the Rebar Is Delivered Where You Want It

Rebar is available at some lumberyards and from most masonry suppliers. However, nine times out of ten, I order it from a steelyard that stocks both grades, that always has #5 bars, and that delivers to the site. If the site has good access and if we have an excavating machine available during the delivery, the machine can lift the rebar from the truck with a chain sling. If we don't have equipment on site, I let my suppliers know so that they can deliver my order on the outside edge of the truck bed, where it can be levered off and onto the ground.

If the delivery is early or late or if the machine is unavailable, we drop the bundles off the side of the truck onto blocks of wood; it's easier to maneuver the chain and slip hooks under the load when we are ready to move it to the foundation area. We also store the rebar off the ground at its staging area so that it doesn't get muddy or sit in puddles. Dirty rebar must be cleaned before use.

The Concrete Reinforcing Steel Institute has something to say about clean rebar in its manual Placing Reinforcing Bars: "The surface condition of reinforcing bars may affect the strength of the bars in bond. The main factors affecting bond are the presence of scale, rust, oil, and mud." Scale is caused by the manufacturing process; loose scale usually falls off when the bar is handled or bent. Tight scale and light rust are acceptable and actually enhance the bond with concrete because they add more surface

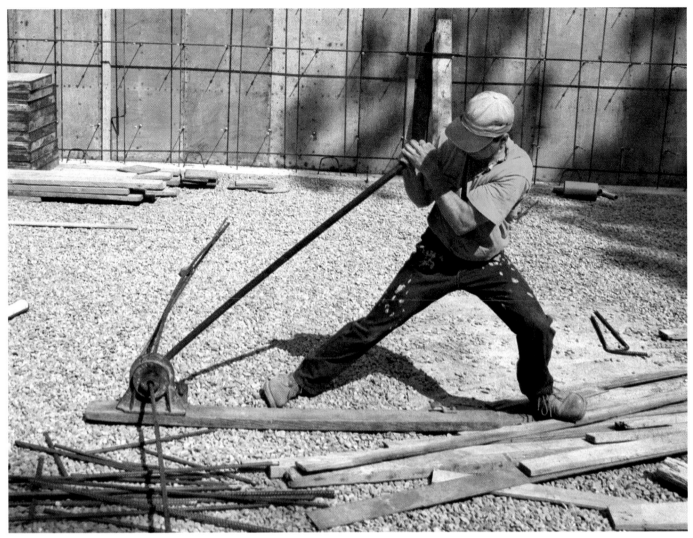

Although chopsaws or torches can make quick work of cutting rebar, at a site without electricity, the stock can be cut with the cutter/bender.

area. Too much form oil on rebar can adversely affect the bond, according to the CRSI manual.

The Grunt Work of Cutting and Bending Rebar

If there's power on site, steel can be cut using a circular saw outfitted with a metal abrasive wheel or a 14-in. chopsaw with the same blade. The latter cuts the bar more easily, but we need to bring the steel to the saw. Using either saw, we can cut three to six pieces of rebar to common lengths simultaneously. Incidentally, wearing goggles and earplugs is a must when cutting steel with a saw.

Without electricity, rebar can be cut to length with an oxyacetylene torch or a cutter/bender (see the photo above). Lacking a torch, we use the slower cutter/bender, mounted on a 2x8 for stability, and cut one piece at a time. Available from masonry-supply houses and some tool catalogs, cutter/benders have cast-steel heads, feature replaceable cutters, and cost $250 to $300. Many cutters have a 52-in.-long handle that gives you plenty of leverage. Even so, the process is slow, and if we're cutting #5, grade-60 bar, we really have to throw our weight into it.

TIP

The best way to mark steel for cutting is with a quick shot of spray paint from a can designed to work upside down.

We've found that the best way to mark steel for cutting is with a quick shot of spray paint from a can designed to work upside down. With anything else, the marks can be pretty hard to see. Exact length is seldom critical with rebar. Because it's designed to overlap, if you're off by an inch, it usually doesn't matter.

The easiest way to bend rebar is with the cutter/bender (see the photo on p. 17). After measuring and marking bends with paint, we lay the stock under the head and lever it into shape, using the tool head as a fulcrum. The stock has to be in a straight line when bent. Although some of the newer models have stops at 90 degrees and 180 degrees, we just eyeball the bend and adjust as needed. We're careful not to overbend an angle, though, because you can't use the bender to bend it back and because it's hard to do using your hands and feet.

To save time, we'll often bend corners while we're waiting for the forms to be completed. If we need more than one bend in a piece of rebar, we've found it's easier to lap with another bent bar; it's awkward to make multiple bends on site, especially with long lengths. In a pinch (such as when we don't have the cutter/bender), we have bent bars around the pintle on my truck's hitch plate, but I wouldn't want to do more than a few bends this way.

Wiring the Rebar Together until the Pour

To lap or cross rebar, we use prelooped wire ties (see the photos on the facing page). Commonly available in lengths of 6 in. and 8 in., the latter are handy for tying together pairs of #4 or #5 bars. A bag of 5,000 8-in. ties costs about $45 and will last for several residential-size foundations. Wire is also available on spools, but we find that less convenient.

The simplest tie requires merely bending the tie diagonally over the bars and, using a tool called a twister (see the top left photo on p. 21), hooking the loops and spinning the

Spray paint marks the spot. A spray can that's made to shoot upside down marks the position of bends or cuts on a piece of rebar. The mark is waterproof, easy to read, and nearly indelible.

Prelooped wire ties are twisted around the junction of two pieces of rebar to hold them in place until the concrete is poured. The wooden-handled hook (top photo), known as a twister, speeds up the process. On longer runs or in place of multiple bends, rebar can be lapped (bottom photo). The overlap should be a minimum of 36 bar diameters and tied together with wire in two places (photo, right).

Keeping Rebar at a Consistent Height

Rebar placed in footings should always be above grade. To prevent the bars from sinking in the wet concrete, foundation crews often suspend rebar from strapping (see the photo at right) that spans the tops of the forms. Any exposed wire is trimmed after the concrete sets up and is covered by the foundation wall.

Manufactured supports known as chairs (see the photos below) offer another option for locating rebar. Available in different sizes, chairs are usually made of plastic or heavy-gauge wire; plastic feet help to insulate the wire from corrosion that results from ground contact.

wire. Overtwisting the wire will simply break it. (The wire adds no strength or integrity once the concrete has been placed.) The twisted wire is then wrapped around the bar (see the bottom left photo on p. 21) so that it doesn't extend toward the exterior surface of the concrete. The wire might rust if it remains exposed to the elements or could lead water into the embedded rebar if exposed below grade. Incidentally, spools of wire are also handy for hanging the rebar at a consistent height in the footings (see the right photo on p. 21). Because the footing is covered later by the waterproofed foundation wall, the wire is never exposed to the elements.

Rebar Maintains Control of Concrete Shrinkage

In addition to solving problems related to shear or tension, rebar is also specified for shrinkage control of concrete (see the drawing below). Because the water in poured concrete is lost by evaporation as it cures, concrete shrinks in volume. Rebar doesn't

TIP

Spools of wire are handy for hanging the rebar at a consistent height in the footings. Because the footing is covered later by the waterproofed foundation wall, the wire is never exposed to the elements.

Rebar Makes the Average Foundation Better

In a typical situation, rebar is placed in a foundation wall to prevent cracking and to strengthen the wall. The exact type and placement of the rebar varies from job to job and is always determined by the structural engineer.

On the basement side of the foundation, rebar should be covered by a minimum of 1½ in. of concrete.

Rebar dowels pin new work to existing structures.

Existing foundation

Verticals tied to horizontals

Horizontal bars minimize cracking.

Rubber caps prevent injury by insulating the sharp ends prior to the pour.

Eighteen-inch dowels of #5 rebar drilled into the ledge are tied to the footing rebar and hold the concrete to the ledge's contour.

Ledge

There should be a minimum of 3 in. of concrete between the soil and footing rebar.

Rebar reinforces footings poured over soft areas in grade.

U-shaped dowels replace keyways. Planted on 4-ft. centers, the dowels help to anchor the footings to the walls and to the rebar.

Rules for Placing Rebar

There are some general rules about placing rebar that can be applied to most situations:

- In footings and walls below grade, rebar should be covered with at least 3 in. of concrete to protect it from groundwater and soil.
- Inside a wall, rebar is positioned on the tension or inner (basement) side with approximately 1½ in. of concrete cover.
- The horizontal shrinkage-control steel is placed first in the wall, followed by shorter vertical bars placed to the inner unrestrained side of the wall.

One note: If there is a dense mat of steel in a slab or wall, it may be necessary to vibrate the concrete with a pencil vibrator to consolidate the concrete without leaving voids.

prevent shrinkage but binds both sides of the eventual cracks into a single wall plane.

According to Val Prest, a structural engineer from Cambridge, Mass., "This shrinkage will result in ⅟₁₆-in. to ⅛-in. cracks at about every 20 ft. at the top of an unreinforced wall. Add more water to the concrete, and you get more shrinkage—cracks that are perhaps every 10 ft. to 15 ft.—and the concrete is weaker. Horizontal bars will minimize the cracking by causing multiple fine line cracks instead."

Prest also says, "Temperature and shrinkage-designed steel for the average 10-in.-thick residential foundation is commonly spaced on 12-in. centers, or every 10 in. horizontally for #4 bars and every 15 in. for #5 bars." For horizontal placement in wall forms, we almost always tie rebar to the tie rods (or form ties) that hold the outer and inner concrete formwork together. After the concrete subcontractor sets up the outer wall and pushes the tie rods through, we place the steel on these tie rods and wire the rebar to them. The verticals are tied to the horizontal bar and to U-shaped dowels in the footing (see the photo on p. 23).

Although steel is specified for its temperature and shrinkage control, Prest says, "Ninety-nine percent of the time, steel is designed for shrinkage, not temperature." Some exceptions would be bridges, concrete roadbeds, and large retaining walls built on highways exposed to direct sun.

Howard Stein is a builder living in Townsend, Massachusetts.

Placing a Concrete Driveway

■ BY ROCKY R. GEANS

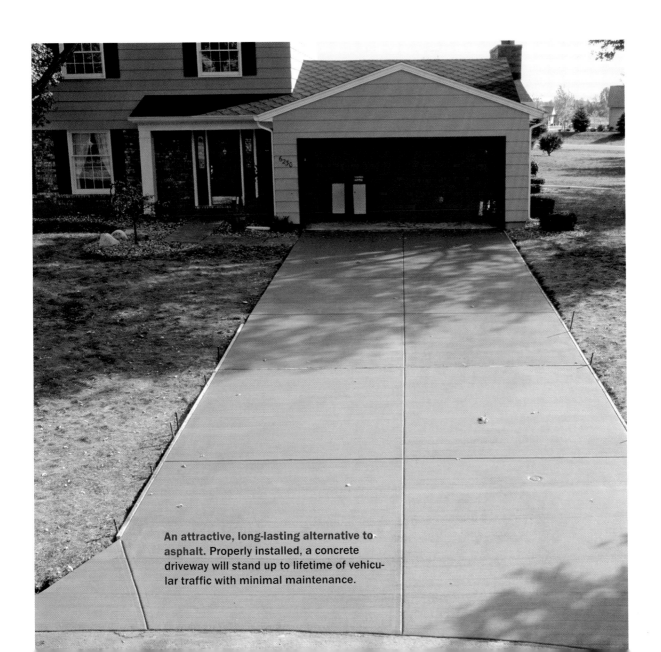

An attractive, long-lasting alternative to asphalt. Properly installed, a concrete driveway will stand up to lifetime of vehicular traffic with minimal maintenance.

Soils Support the Driveway

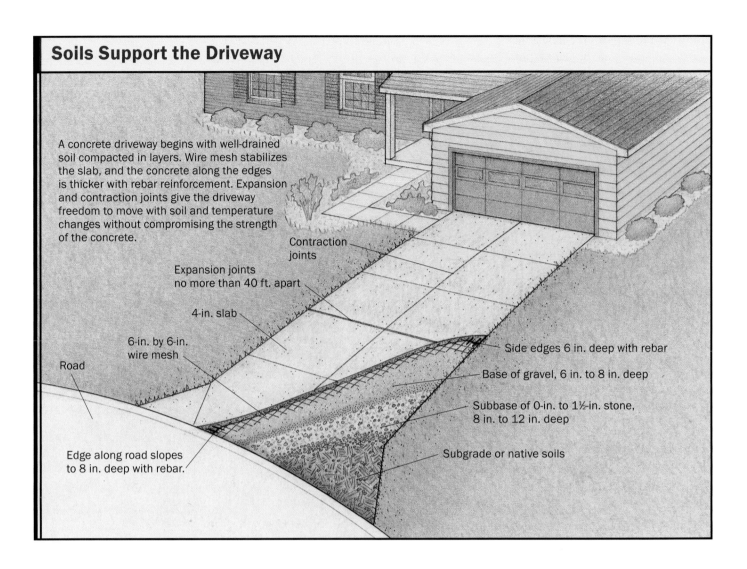

A concrete driveway begins with well-drained soil compacted in layers. Wire mesh stabilizes the slab, and the concrete along the edges is thicker with rebar reinforcement. Expansion and contraction joints give the driveway freedom to move with soil and temperature changes without compromising the strength of the concrete.

Contraction joints

Expansion joints no more than 40 ft. apart

4-in. slab

6-in. by 6-in. wire mesh

Road

Edge along road slopes to 8 in. deep with rebar.

Side edges 6 in. deep with rebar

Base of gravel, 6 in. to 8 in. deep

Subbase of 0-in. to 1½-in. stone, 8 in. to 12 in. deep

Subgrade or native soils

A crucial part of proper driveway design is making sure the materials below the concrete are adequate.

A few years back, we had just finished a driveway at a house in South Bend, Indiana. We were about to call it a day when the homeowners' dog got loose, burst into the garage, and used our freshly placed concrete drive as his escape route. When the dog heard us holler, he stopped in the middle of the driveway. Then we tried to get him back inside, which made him scamper back and forth on top of the concrete, making a bad situation worse. We ended up refinishing the whole driveway and giving that dog a scrubbing he'll never forget.

A Driveway Is Only as Good as the Soils Beneath It

We've been installing concrete driveways for 30 years, and a crucial part of proper driveway design is making sure the materials below the concrete are adequate (see the drawing above). The first 6 in. to 8 in. of material directly below the concrete is the base. The subbase is the soil 8 in. to 12 in. below the base, and the subgrade is usually the native or naturally occurring soil below the subbase. The design thickness of each layer depends on the soil being built on. Acceptable natural soils such as sand and gravel let moisture drain. If the subgrade consists of this type of soil, then it can be compacted to serve as the base and sub-base, and more excavating and filling are unnecessary.

However, if the subgrade is clay, peat, or fine-grained silty soil that holds moisture and drains poorly, removal of up to 20 in. of subgrade soil might be necessary, depending on the support value of the soil. If you have doubts about the soil characteristics in your case, it's worth hiring a soils engineer to do an evaluation.

Establish Driveway Elevations Early for Proper Drainage

Prior to excavating and backfilling, the exact elevation of the top of the drive should be established. Then, as earthwork is being done, base grades can be brought up with equipment usually to within an inch of their required height, which saves on hand-grading later. For the best drainage, we try to slope the driveway at least ¼ in. per running foot away from the house.

Some situations prevent proper drainage, such as an area of concrete that is locked between a house and a garage. In these cases a catch basin may have to be installed as part of the driveway's drainage system.

The best way to remove water from a catch basin is to use a drain pipe at least 4 in. in dia. that returns to daylight or to a storm sewer that is located safely away from the house. A second method is connecting the catch basin to a dry well. In the most extreme cases, a sump pump is installed to pump collected water to a safe place. The last solution is the most costly and probably should be used only with the recommendation of an engineer.

Soils Must Be Compacted to Support the Driveway

I check soil compaction initially by walking over the area to get a feel for firmness. Additionally, I shove a ¼-in.- to ½-in.-dia. smooth

TIP

The best way to remove water from a catch basin is to use a drain pipe at least 4 in. in dia. that returns to daylight or to a storm sewer that is located safely away from the house.

Recycled concrete forms the subbase. Coarsely crushed concrete is compacted to provide a drainage layer and a capillary break to prevent moisture from wicking up.

steel rod into the soil in several places to check the resistance. If the soils are properly compacted, the rod should encounter firm, even resistance over 2 ft. to 3 ft. However, if the rod meets resistance, say, in the first 6 in. and then can be pushed farther into the soil with ease, it's a sure indication that only the top 6 in. of soil is compacted and that the lower layers of soil are loose. Over time, loose, uncompacted soils will settle as storm water drains through them. Soil settling leaves voids that greatly increase the odds for driveway cracking, sinking, or even collapsing in certain areas.

If testing reveals uncompacted soil layers below a top layer that is compacted, then the top soils need to be removed and lower levels compacted properly before the base layers are replaced. In many instances the top 12 in. may be properly compacted while the next 3 ft. to 4 ft. are loose. Ideally, the soils should be excavated down to solid native soil (usually no more than 4 ft.). The soil is then replaced and compacted in 6-in. layers called lifts.

In new construction where excavation has occurred for a foundation, soils should be backfilled and compacted in 6-in. lifts all of the way to final grade. Otherwise, concrete work such as driveways, sidewalks, and patios will settle over time and slope or fall inward toward the house. If there are doubts about compaction in any situation, some companies perform on-site compaction tests, a minor investment that can buy peace of mind.

In addition to being properly compacted, the subbase and base need to be made out of materials that will form a capillary break. A capillary break is a layer of soils large and coarse enough to prevent water from being drawn up into them through capillary action the way lamp oil is drawn through a wick. Moisture allowed to wick up and accumulate in soils beneath the driveway slab will freeze, expand, and create frost heaves in the slab. Zero-inch to 1 ½-in. stone or re-

cycled concrete for a subbase compacts well and creates a good capillary break.

Forms Determine the Final Grade of the Driveway

After the soils have been layered and compacted satisfactorily and the proper grades established, the driveway forms can be laid out and installed. We usually make our forms out of 2x4s if the driveway is to get a 4-in. slab or 2x6s for a 6-in. slab. The forms need to be staked strongly enough to hold the concrete and to withstand screeding without movement, so we drive our stakes every few feet. Wooden stakes are okay, but they don't hold up well to being driven and generally can be used only a few times. On most jobs we use commercially available round metal stakes with predrilled holes for nailing the stake to the form.

The driveway featured here was placed on a nice, gently sloping lot. We began by setting the forms to the natural grade along one side. On the opposite side we ran a string at the exact width of the finished slab. We staked the forms for this side along the string at roughly the correct height. Next, we leveled from one side to the other to set the exact height of the opposite forms, driving the stakes deeper or pulling them up to adjust the height. We've found that a laser level is the quickest and easiest tool for setting the height of our forms, although a transit or even a water level can be used.

Our forms usually receive a coating of release agent (a special oil available from concrete-supply houses) to provide better consolidation of the adjacent concrete and to make removal easier. After the forms are set, we grade the base layer of soil to its exact elevation (see the top photo on the facing page) and run the compactor over it one last time. Then we dig the edges of the driveway down a couple of inches. The thicker concrete along the perimeter provides additional support as well as protecting against

The base layer is graded to the proper height. The laser level in the background is used to set the forms to the correct height. The crew then grades the base to the forms, and the base gets compacted one last time.

erosion of the soils next to the drive. We incorporate a single run of rebar along the edges for additional support, as well as an 8-in.-thick by 12-in.-wide thickened area at the road.

Expansion Joints Allow the Concrete Slab to Move with Changes in Weather

Another crucial part of the driveway layout is planning for contraction and expansion joints. Contraction joints are added during placement, so I'll discuss them later. Expansion joints, installed before concrete placement, allow the driveway to move both horizontally and vertically. Most people think concrete is solid and unmoving. However, concrete not only moves in relation to other solid structures, such as foundations and roadways, but also expands and contracts with temperature changes and moves as soil conditions beneath the slab change.

Expansion joints provide a full division between different sections of concrete placement. For the driveway featured in this article, expansion joints were placed between the driveway and the sidewalk to the front door, between the driveway and the garage apron, and between the two main-driveway slabs that were placed or poured separately. We didn't need an expansion

Expansion joints get a plastic cap for installation. Expansion joints act as buffers between driveway sections. After the concrete has cured, the plastic cap is removed, and sealer fills the joint to keep out moisture and light.

joint between the driveway and the asphalt roadway, but a joint is required if the roadway is concrete.

An expansion joint consists of a thin layer of energy-absorbing material such as asphalt-impregnated fiberboard, plastic foam, wood, cork, or rubber. For most driveways, we use ½-in.-thick fiberboard installed with a plastic cap strip on top (see the photo above), flush with the finished height of the slab. After the concrete has set up, the cap strip is removed, and we fill the top of the joint with a joint sealant (SL-1 from Sonneborn, Chem-Rex Inc.), which protects the joint from moisture penetration and UV-degradation. The joint sealant also matches the color of the concrete to add a more pleasing look to the joint.

Expansion joints serve several functions in a concrete driveway. First, the expansion joint provides relief between slab sections as the concrete expands and contracts with temperature changes. This movement is horizontal, and the joints that serve this function should be placed no more than 40 ft. apart in any direction. To facilitate this horizontal movement, the base that rests below the slab has to be smooth and well compacted, with no obstructions such as rocks or holes, which can fill with concrete and restrict movement.

The concrete slab needs to be able to slide back and forth over the base. Any restriction of this movement will contribute to cracking. Even the thickened edge at the roadside is designed with a gradual slope from 4 in. to 8 in. rather than having an abrupt 90 degree excavation that would restrict slab movement.

Expansion joints also serve as a buffer between the driveway slab and the adjoining rigid structures, in this case the sidewalk and the garage. Movement at these points is vertical; the driveway moves in reaction to changes in the soils beneath the slab. An example is the soil swelling that occurs when moisture in the slab freezes. Without the aid of the expansion joint, the slab would chip or crack as it slid by slight imperfections in the abutting concrete.

Another way we restrict vertical movement is by drilling and installing steel dowels into adjoining rigid cement work, such as a foundation just below the driveway slab. A ⅜-in. by 12-in. dowel works well in most situations.

The foundation, which is bearing on a footing below frost level, should not move. However, the driveway, which rests on soils only 4 in. to 6 in. below grade, will almost certainly experience some degree of movement. We want to give the drive the freedom to move up but not to drop any lower than the dowel.

We also use dowels to maintain alignment between the driveway and existing flat work, such as sidewalks. The dowel should be smooth and installed parallel to the concrete, not at an angle. Rebar can be used for this application, but it won't work as well because of its rough texture. We drill a hole into the existing flat work slightly larger than the diameter of the dowel, which allows for expansion and contraction. These dowels should be installed a minimum of 2 ft. o.c.

Wire Mesh Is the Best Reinforcement Option

Next we place the reinforcement. We use 6-in. by 6-in. wire mesh throughout the slab as reinforcement. Wire mesh is not designed to prevent concrete from cracking. However, it does prevent widening or horizontal separation of cracks that form.

There are many claims that cracking can be eliminated by mixing nylon or steel fibers with the concrete. I've used fiber mesh on a couple of large parking lots, and I have mixed feelings about the long-term results. I still think that wire-mesh reinforcement installed correctly is best.

In most cases wire mesh is pulled up into the wet concrete while it's being poured, ideally 1 in. to 1½ in. above the bottom of the driveway slab. We have the concrete truck place a bead of concrete at a uniform height that supports the wire mesh while the rest of the concrete is being poured (see the photo on the facing page). Another

Wire mesh is pulled up on top of a thin layer of concrete. A crew member pulls the wire-mesh reinforcement up to the proper height with a rebar hook. A bead of concrete laid first supports the mesh during placement.

method that I prefer (although it's probably more time-consuming) is setting the wire mesh on 1½-in. chairs that keep the wire at a more consistent height.

When the wire mesh is in place, I make sure everything has been prepared properly and that every tool needed for finishing the slab is on hand. The next step is ordering the concrete.

Order the Right Concrete Mix

Concrete mixes vary depending on the application. But for most driveways, we use a six-bag limestone mix (4,000 psi) with approximately 6 percent air entrainment. Air entrainment is the incorporation of microscopic air bubbles throughout concrete to help prevent scaling, the flaking or peeling that occurs on cured-concrete surfaces.

The mix order should also include the slump requirement, or the wetness of the concrete when it's delivered. Slump is measured on a scale of 1 to 12, with 1 being the driest mix. For most driveways, we request a slump of 4 to 5, which is easy to spread

but can be worked shortly after it is poured. After a mix has been prepared to specifications, adding water can weaken it. The concrete supplier is responsible for the slump as well as the strength of the mix, and the concrete should arrive as ordered. If concrete arrives too wet, it can be sent back.

Screeding Creates the Level of the Slab

Trucks in our area are able to distribute concrete pretty evenly by controlling the flow of material and the direction of the chute. We work the concrete along the edges to consolidate it and to remove any voids. Then the concrete is raked to a rough elevation just slightly higher than the forms and expansion joints. We make sure we never get too far ahead of the screeding process so that any excess can be easily raked down to areas waiting for concrete.

Screeding cuts in the grade of the slab and consolidates the concrete before bull floating (see the top right photo on p. 32). It's usually done with a long, straight 2x4 (we sometimes use an aluminum box beam

The best bull floats have a blade that rotates back and forth by simply twisting the handle, allowing the operator to keep the handle at a constant, comfortable angle.

Working the edges. A crew member with a flanged edging tool works along the perimeter of the slab to consolidate material and to rough-cut the rounded edge.

Screeding cuts concrete to the right height. A long, straight 2x4 or aluminum box beam is pulled across the surface of the concrete while it is moved in a side-to-side motion. Screeding brings the level of the concrete down to the top of the forms and consolidates it at the same time.

A bull float smooths the concrete and fills voids. As the crew member in the foreground pulls the wide blade of the bull float over the concrete, surface tension is created to bring water to the surface and fill in imperfections from screeding.

screed rail), slightly longer than the width of the driveway. The 2x4 or screed rail that rides on the forms is pulled across the wet concrete with a side-to-side reciprocating motion.

After two or three yards have been placed, the edges should be hand-floated and cut in with an edging tool (see the top left photo). The screeded concrete can now be bull-floated (see the photo at left). A bull float is a wide, flat metal float mounted on the end of a long handle. As the float is pushed and pulled over the screeded concrete, the leading edge must be elevated to keep from digging in. If the float is mounted on the handle at a fixed angle, bull-floating can be a real workout. The best bull floats have a blade that rotates back and forth by simply twisting the handle; this design allows the operator to keep the handle at a constant, comfortable angle. As the bull float rides over the concrete, it creates surface tension that brings water to the top, which smooths the slab and fills in minor voids at the same time.

After bull-floating, the concrete should be left alone until all bleed water on top of the wet concrete has evaporated. At this point the concrete should be strong enough to support a crew member on kneeboards, which I'll describe later, and is ready for finishing. Finishing the concrete too early can trap water and create a weak surface with a high water/cement ratio.

Contraction Joints Give the Slab a Place to Crack

The first part of the finishing process is laying out and cutting the contraction or control joints (see the photo below). Contraction joints act as score marks in the concrete; they create weak points and encourage any cracks that might develop to occur at the joints. To understand how contraction joints work, I need to explain why concrete cracks.

Concrete begins to crack before receiving any loads whatsoever. As the concrete cures and dries, water is absorbed into base materials and evaporates through the surface, which causes the concrete to shrink or contract. Cracks form in the concrete as a result. Contraction joints provide the relief needed so that these cracks form along a joint instead of randomly in the surface of the slab.

Maximum spacing of contraction joints should follow this rule of thumb. Multiply the thickness of the slab by 2½, and that number represents the maximum distance in feet between joints in any direction. The slab for this driveway was 4 in. thick, so the maximum distance between the joints is 10 ft. (4 x 2.5 = 10).

The depth of the joint should be no less than one-quarter of the thickness of the slab and should be cut in either during the finishing process or immediately afterward. For this driveway we cut the joints with special tools called groovers. We begin by stretching a string between our layout lines and snapping it to leave an impression in the wet concrete. We work the groovers along straightedges to cut in the joints. Thicker slabs require deeper joints that are cut with saws equipped with blades designed to handle fresh, or green, concrete.

TIP

Maximum spacing of contraction joints should follow this rule of thumb. Multiply the thickness of the slab by 2½, and that number represents the maximum distance in feet between joints in any direction.

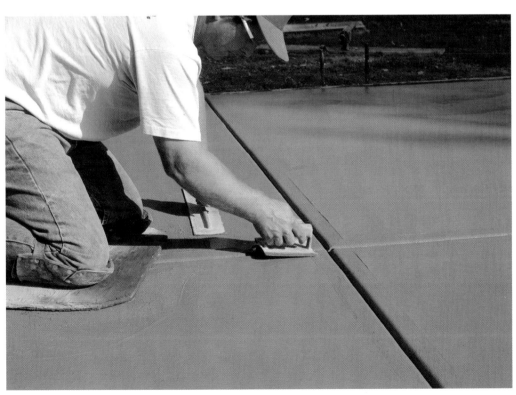

A special tool cuts the contraction joints. Contraction joints score the slab so that cracking from the curing process occurs at these points rather than at random.

TIP

Make kneeboards out of ½-in. plywood, about 24 in. long and 16 in. wide. Cutting off the corners will keep them from digging into the fresh concrete. In addition, a handle on one end will help a lot when a crew member is moving them around on the slab.

Concrete treated with curing compound has better resistance to scaling.

Kneeboards keep the middleman from sinking in the concrete. The crew works the surface for the last time, going over it first with magnesium floats and then doing a final smoothing with flat-bladed trowels. The crew member in the middle works on plywood boards that distribute his weight and keep him from sinking into the slab.

Kneeboards Distribute Body Weight When You're Finishing Fresh Concrete

After bull-floating, all of the steps in finishing concrete, including cutting in the contraction joints, require a crew member to work in the middle of the slab on top of the uncured concrete. To keep from sinking into the fresh concrete, this crew member works on a pair of kneeboards (see the photo above). We make our kneeboards out of ½-in. plywood about 24 in. long and 16 in. wide. We cut the corners off to keep them from digging into the fresh concrete, and we put a handle on one end that helps a lot when the crew member is moving them around on the slab.

After the contraction joints have been cut, the crew works the surface of the slab one last time. For most driveways one crew member works the center of the slab while two work the edges. First they go over the surface with an aluminum or magnesium hand float with a thick blade slightly round in section. This process, known as magging, releases air that might be trapped in the concrete from bull-floating and leaves a smooth, open texture on the uncured concrete.

As a final step, a trowel with a thin, broad metal blade is passed over the surface in large circular strokes. Troweling should leave the surface of the slab smooth and flat. If any slurry from the magging gets into the contraction joints, it will be necessary to go back over them with the groover and blend the edges of the groove into the rest of the slab with a trowel.

A broom finish provides traction. A broom finish is applied to the still-wet concrete after it is troweled. A wide, fine-bristled broom is dragged slowly in parallel strokes from the middle of the slab out and is cleaned with water between strokes.

A Broom Gives the Driveway a Rough Surface

We give most of our exterior flat work, including driveways and sidewalks, a broom finish for traction. Right after troweling, a crew member drags a wide broom over the slab in smooth, parallel strokes (see the photo above). We use a fine-bristled broom made either of nylon or of horsehair. A coarse broom will dig into the surface too deeply and dislodge the aggregate. Because this driveway was double wide, the crew member started at the middle and dragged the broom to the outer edge for each side of the slab. The broom should be cleaned by dipping it in water after each stroke to keep excess concrete from building up in the bristles and changing the texture.

Right after we finish, we spray on Kure-N-Seal® (made by Sonneborn), a combination curing and sealing compound. Used to prevent water from evaporating too rapidly, curing compounds form a membrane on the surface of a slab. Application of curing compound effectively slows the curing rate, and the longer concrete takes to cure, the stronger it becomes. Concrete treated with curing compound also has better resistance to scaling. Sealing compounds prohibit moisture from getting into the concrete once the concrete has cured. Because the concrete's curing takes several days, we recommend not allowing vehicular traffic on the slab for a week.

Rocky R. Geans is the owner of L. L. Geans Construction Co. of Mishawaka, Indiana.

Drilling Concrete? Get a Rotary Hammer

■ BY DAVID CROSBY

Concrete is a versatile but unforgiving material. The same qualities that make it so useful also make concrete tough to work after it has cured. The most common way that builders, remodelers, and carpenters encounter cured concrete is when they need to put holes into it, such as for anchor bolts, pins, dowels, and plumbing or electrical chases.

You could make holes the old-fashioned way, with a hammer and a star drill—a piece of hardened-steel that's struck, then rotated, struck, then rotated—to get a hole, eventually. Or you could use a hammer drill (see "How Does It Work?", on the facing page) for a few small holes. But for quickly drilling a lot of holes in concrete or in a masonry wall, nothing beats a rotary hammer.

Look out, though: With competition fierce and manufacturers making a broad range of rotary hammers to fill subtle niches, the number of choices available can be overwhelming. To make a first pass at limiting your search, ask yourself what size holes you intend to drill.

SDS shank is held in place by bearings.

Carbide-steel drill bit rotates 800 to 900 times per minute while delivering 3,000 to 4,000 impacts per minute.

How Does It Work?

Old-timers made holes in concrete with a star drill, a piece of hardened steel with a star-shaped chisel on the end, that was struck with a big hammer, rotated an eighth of a turn, then struck again. Eventually, you would get a hole.

The piston compresses air within the pneumatic chamber, driving the bit forward. On the return stroke, the vacuum created pulls back the bit, preparing it for the next impact.

The dropped-motor style is geared so that it places the motor at a right angle to the bit. This compact configuration favors larger motors that deliver rotation and impact without adding length.

Rotary Hammer Drills

With a rotary hammer, the bit pounds and turns at the same time, but the similarity ends there. There is no hammer delivering the impact. Instead, a gear-driven crank moves a piston back and forth within the pneumatic chamber. The compressed air created by this motion propels the carbide-steel bit forward, delivering concrete-pulverizing impact. Drilling into 3000-psi concrete, this engineering marvel bores ¾-in. holes, 4 in. deep, in about 30 seconds.

Hammer Drills Are Different

Hammer drills, which are lighter than rotary hammers, tap away material with a percussive mechanism like two poker chips rubbing together. If you push harder (within reason) on a hammer drill, you bore faster. The opposite is true of rotary hammers. Their bits actually move back and forth, or reciprocate, with each blow. If you lean in to a rotary hammer, you only slow it down and wear it out sooner. Just hold it securely and let it do the work.

THREE TOOLS IN ONE

Rotary hammers typically operate in three modes: hammer only (chipping), rotary/hammer, and rotary only (drilling). Tools in the 1-in. to 1¼ -in. range are well suited for residential work, not just because they bore holes in concrete quickly, but when chipping concrete they're powerful enough to be useful. In drilling mode, these tools bore large holes in wood easily.

Start by Choosing the Right-Size Tool for the Job

Rotary hammers fall into two general size classes: The smaller ones accept bits with SDS or SDS-plus shanks, and the larger hammers use either SDS-max (looks similar to a larger SDS bit) or spline drive (looks like the spline on a drive shaft or truck axle). Within these classes, the tools are categorized by their nominal drilling capacity, or the largest hole that the tool is designed to drill efficiently.

Even though rotary hammers are available with drilling capacities from ¾ in. to 2 in., 85 percent of all holes drilled in residential construction are ¾ in. or smaller. This is a good point of departure for narrowing the field, but there's more to the story.

Constantly operating a power tool at full capacity may fall within the manufacturer's specifications, but over the long haul, you, the tool, or both may suffer early burnout. A little

THE TEST

To be objective, I built a rotary-hammer drill press. Each tool was outfitted with a new ¾-in. carbide-masonry bit prior to drilling five consecutive 4-in.-deep holes (gauged with a stop block). The times for each drill were recorded, then averaged, and are noted in the comments for each tool. Surprisingly, all the tools drilled faster in the drilling jig than they did handheld.

BOSCH® 11236VS.

A first-rate machine, the Bosch is well suited for professional use. I found no problems with it in any regard. The case is by far the best. It permits easy, well-organized storage for tools and bits. The depth stop is well made, and chisels can be rotated to 36 different positions. The controls are easy to access, and the overall quality is good. It isn't as powerful as the Makita or the Hitachi, but the Bosch is a good buy for the money.

Capacity	1⅛ in.
Weight	9.3 lb.
Drilling test (sec.)	32
Cost	$300

Once You've Drilled the Holes: Concrete Anchors

Lead shields and lag screws are commonly used to anchor lumber to masonry.

Tapcon screws, made from epoxy-coated, hardened steel, thread directly into concrete. They're easy to remove and replace. Lead shields aren't necessary.

extra power goes a long way toward improving comfort, ease of use, and productivity.

Of course, too much of a good thing shows up as extra size, weight, and cost. The happy medium for most residential work falls within the 1-in. to 1¼-in. range of drilling capacities. These rotary hammers, at the upper end of the SDS category, drive ¾-in. bits smoothly without working too hard. All the hole-drilling tests for this review were done with ¾-in. bits.

Don't Underestimate Their Versatility

Even though 1-in. to 1¼ -in. rotary hammers generally are used for boring concrete, they also have rotary-only (drilling) and hammer-only functions. What sets this class apart is that unlike smaller rotary hammers, the hammer-only function on the 1-in. to 1¼-in. models is powerful enough to be useful. This size is great for light demolition, removing tile, brick, block, and small amounts of concrete, especially in tight spaces. In

DEWALT® D25304K.

The DeWalt is light (6.8 lb.), but with rotary hammers, light weight is not always an advantage. Its compact length (11.5 in.) is definitely an advantage in tight places. The D25304K was loud and appeared to labor as it drilled in the hole-boring test. It was easy to drill too deep with the plastic depth stop.

Capacity	1 in.
Weight	6.8 lb.
Drilling test (sec.)	36
Cost	$350

Includes a quick-change removable SDS chuck and a ½-in. three-jaw chuck.

HITACHI® DH30PC.

This tool is a fine machine, a smooth and powerful performer, second only to the Makita. It's compact in size and has the best handle and depth stop in the group. Vibration and noise were minimal when drilling holes or during light chipping. The Hitachi comes with a dust collector for overhead drilling. My only complaints are that the switch for selecting modes of operation is undersize for gloved hands, and the case is so small as to be troublesome.

Capacity	1¼ in.
Weight	10.6 lb.
Drilling test (sec.)	26
Cost	369

Wedge anchors are medium- to heavy-duty anchors. The material being fastened can be used as a template for the hole being drilled; both holes are the same size.

Strike anchors, similar to wedge anchors, are set easily by hammering a steel pin, which spreads the anchor's flanges. They are difficult to remove.

drilling mode, these rotary hammers also can be used to bore large holes in wood.

Contributing to the extra power and compact size is the squarish dropped-motor style common to all the tools, except the Metabo. Orienting the motor in this way puts it at a right angle to the bit, allowing the tools to be designed with larger motors but without added length. That said, these rotary hammers remain light enough to handle; each hammer weighs in at less than 11 lb. None of these tools is prohibitively expensive.

Each of the rotary hammers features variable speed, but the size and placement of other controls vary by manufacturer. Using these tools is hard work, and I'm in favor of any design that makes it easier to adjust the controls or to change bits with a gloved hand.

The six hammers that I tested each came with a depth stop and a case. Some of the cases stored bits and accessories better than others. Although this may seem like nitpicking, remember that a rotary hammer is likely to be one of the more expensive power

MAKITA® HR3000C.

The Makita is my favorite rotary hammer in the group. Its heavier weight and slightly larger capacity proved to be advantages. The controls are easy for gloved hands to use, and the variable-speed switch is accurate, steady, and predictable. This tool features a clutch, which disengages the motor from the output shaft in case the bit binds during drilling. The long cord with strain relief and the easy-to-use chuck for speedy bit changing are great user-friendly features. The generously sized case could be designed more efficiently.

Capacity	$1\frac{3}{16}$ in.
Weight	10.1 lb.
Drilling test (sec.)	18
Cost	$500

Concrete Anchors Continued

Drop-in anchors lie flush to the surface of the concrete. Machine screws thread into them.

Sleeve anchors, similar to wedge anchors, are light- to medium-duty fasteners.

tools on the job. Bits and accessories easily can equal the cost of the tool. At the end of a cold, muddy day, cleanup and bit storage have to be easy. If it's hard to stay organized, chances are you won't.

Narrowing the Field by Cost

Features are important, but a purchase comes down to this: How much tool can I get for my money? Many power tools used to build houses can be reliably (if artificially) divided into three groups: less than $200, less than $400, and less than $900. As a professional builder doing residential work, I wouldn't spend $900 for a drill. But I'd definitely spend $200, and if I could get a tool that would do a lot more work for around $400, I'd seriously consider it.

I tried to limit the field of 1-in. to 1¼-in. rotary hammers to less than $400. This size offers enough power to perform residential tasks well at a price that's not prohibitive. The price range worked—until I considered tools by Makita and Hilti®. In the case of

METABO® UHE 28MULTI.
The Metabo is a light, well-balanced, finely crafted tool. It has a "smart" switch, which prevents the tool from restarting if power is temporarily interrupted while the switch is locked on. It was pleasant to use and quiet, and it had relatively low vibration. Even though the Metabo had no apparent deficiencies, it did not deliver performance equal to the other tools in the group.

Capacity	1³/₃₂ in.
Weight	6.2 lb.
Drilling test (sec.)	37
Cost	$345

Includes an SDS chuck and a keyless ½-in. three-jaw chuck.

MILWAUKEE® 5359-21.
This tool performed well in the timed drilling test. The side handle is easy to attach and remove, and the tool features an easy-to-use depth stop that adjusts with a lever. Similar to the Makita, this tool has a "slip clutch" that disengages the rotary action if the bit binds during drilling. This is a great feature and should be standard on all large drills. The 5359-21 is a solid tool; I found nothing about it to dislike.

Capacity	1⅛ in.
Weight	8.3 lb.
Drilling test (sec.)	27
Cost	$345

Double-expansion anchors are similar to drop-ins. Their lead, top, and bottom wedges expand as the machine screw is tightened.

Rebar set in epoxy is commonly used for concrete dowels.

SDS Bits Are for More Than Drilling Holes in Concrete

Rotary hammers excel at boring concrete, but their drilling/reciprocating action can be put to work in other ways. For light concrete demo, chisels or points are useful. Core bits are available for drilling holes in concrete up to 3 in. in dia., and for amazingly fast rough-wood chiseling, Bosch offers a set of wood chisels.

SDS shank

Drill bit

Point

Chisel

Core bit

Wood chisel

Makita, I included the company's comparable rotary hammer, the slightly heavier HR3000C, because it wasn't far out of the price range (around $500). But at about $625, Hilti's TE 15-C DLX seemed like too much of a stretch, so I didn't include it.

Putting the Rotary Hammers to Work

To get a better picture of how well these tools perform, it was important to note their day-to-day performance as well as to create a more objective hands-off drilling test. The way a tool feels in your hands is subjective, but anyone familiar with power tools knows that often the feel of a tool is a reliable indicator of quality. Better tools look solid, feel solid, and operate powerfully and smoothly. When it comes to rotary hammers, lower vibration translates to less user fatigue.

I studied the owner's manuals, then spent some time drilling lots of ¾-in. holes and chipping chunks off a large block of 3,000-psi concrete, batched with highly consistent ⅜-in. or smaller aggregate (the kind used for filling insulated concrete forms).

In addition, I did some drilling using a homemade drill press (see the far left photo on p. 38), partly to hold the drills in a constant vertical orientation and partly to remove operator inconsistency. Even though all the tools are rated at about 7.5 amps (again, the Makita was rated slightly higher at 8.2 amps), differences were apparent immediately.

Drill Fixture Yields Surprise

When using a rotary hammer, a firm, steady, stable grip is important for several reasons. Foremost is safety: If one of these hammers gets away from you, it could be painful. Second, if you don't hold the rotary hammer still relative to the workpiece, you will end up with an odd-size hole. This may seriously weaken the holding power of an expansion anchor. Third, a firm grip results in greater efficiency. Even knowing this, I was surprised by how much faster these tools worked in the drilling rig I made.

Earlier, when doing handheld drilling using the rotary hammers' depth stops, the tools averaged from 35 to 45 seconds to drill a ¾-in. hole, 4 in. deep. After clamping the tools into the drill fixture, the drilling times were cut nearly in half, with nothing more than gravity and the 6 lb. or so from the steel sleeve aligning the hammer with the jig.

Drilling, both handheld and with the drill fixture, was done on the same chunk of concrete. My hunch is that handheld drilling was slower due to too much pressure from the user, which lowered the tool's efficiency, or due to unintended lateral movement, or a combination thereof.

Two Rotary Hammers Stood Out

If I had to stay within my $400 limit, I'd choose the Hitachi. It has the best depth stop and side handle; I also think it's the best tool for the money. Overall, though, my favorite is the Makita HR3000C. It's well designed and smooth-drilling, and it's in a class by itself when it comes to drilling speed. But at around $500, the Makita is in a higher price class, too.

David Crosby *As a child, David Crosby saw carpenters put a roof on a house and thought construction had to be the best job in the world. After serving in the Air Force, he followed up on his childhood dream, working as a builder in Arizona, Washington, Connecticut, Maine, Maryland, North Carolina, and New Mexico.*

Sources

Bosch
4300 W. Peterson Ave.
Chicago, IL 60646
877-267-2499
www.boschtools.com

Concrete Fastening Systems Inc.
1231 E. 26th St.
Cleveland, OH 44114
888-498-5747
www.confast.com

DeWalt
626 Hanover Pk.
Hampstead, MD 21074
800-433-9258
www.dewalt.com

Hitachi
3950 Steve Reynolds Blvd.
Norcross, GA 30093
800-829-4752
www.hitachi.com/powertools

Makita
14930 Northam St.
La Mirada, CA 90638
800-462-5482
www.makitatools.com

Metabo
1231 Wilson Dr.
West Chester, PA 19380
800-638-2264
www.metabousa.com

Milwaukee
13135 W. Lisbon Rd.
Brookfield, WI 53005
800-729-3879
www.milwaukeetools.com

A Solid Deck Begins with Concrete Piers

■ BY RICK ARNOLD

Dig a hole and fill it with concrete. How hard can that be? I've seen old decks built on top of little more than a shovelful of concrete, cinder blocks up on end, and even 8-in. by 12-in. patio blocks. I've also seen old decks—not to mention a couple of new ones—sink and pull away from a house, heave up with the same results, and even both sink and heave from one end to the other.

An insufficient design or a bad installation of this simple foundation system can have disastrous consequences in terms of safety, aesthetics, and a builder's reputation. That's why I approach piers with the same care as I do a house or addition foundation.

Soil Conditions and Load Requirements Determine Pier Size and Spacing

Because piers perform the same job for the deck that the foundation does for the house,

it's critical to size and space them properly (see "Everything You Need" on the facing page).

I begin by figuring out how many piers I'm going to need. This decision depends mostly on deck design. For this project, I was building a simple 12-ft. by 16-ft. rectangular deck with a double rim joist to act as a beam that could span about 8 ft., with posts running down from the beam to the piers. In this type of application, I start with two piers on the corners and divide the 16-ft. double rim joist until I get a figure of 6 ft. or less. Here, I found that dividing the rim joist into three sections gave me a span of roughly 5 ft. 4 in., which came out to four piers. Even though my double rim joist could span 8 ft., I chose to use a 5-ft. 4-in. pier spacing to minimize the pier diameter.

After calculating the number of piers I need, I determine the size they need to be. The size of builder's tube dictates the size of the bottom of the pier, which is the area that will be in contact with soil at the bottom of the excavation. To figure this out,

To do so effectively, piers need to be sized and spaced according to the deck's design load and the soil's bearing capacity (see "How Many and What Size?" on p. 48). In cold climates, piers always should sit below the frost line to prevent frost heaves. Check your local code for pier-depth requirements.

Footings Spread the Weight over a Larger Area

If the piers will be used in soil with poor bearing capacity or if the deck has a heavy design load, use a spread footing to distribute the load over a greater surface area. The more expensive, labor-intensive way to do this requires a relatively large excavation for each pier. After forming and pouring the footings, you have to install the tubes and backfill around them, then complete a second pour for the piers. But plastic footing forms bring this process down to just one pour.

For most of these systems, a builder's tube is fastened to the top of the form; then the assembly is lowered into the hole, backfilled, and poured in one shot. See the manufacturer's website for sizing and load requirements.

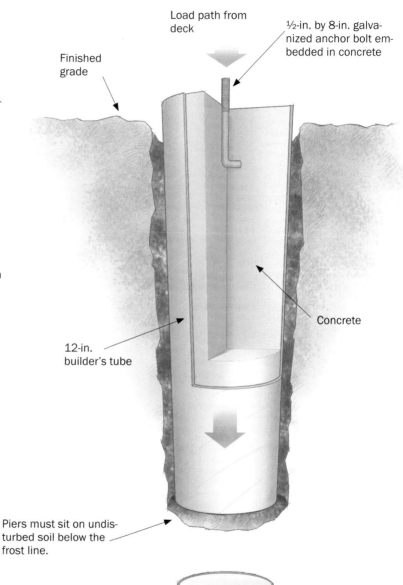

Load path from deck

½-in. by 8-in. galvanized anchor bolt embedded in concrete

Finished grade

Concrete

12-in. builder's tube

Piers must sit on undisturbed soil below the frost line.

When using a spread footing, the load is distributed over a larger surface area than a builder's tube alone can cover.

24-in.-wide spread footing

I calculate the maximum weight each pier must be designed to bear (by code). For this deck, I figured a 1,600-lb. load on each of the two inside piers (see "How Many and What Size" on p. 48). Then I compared that to the bearing capacity of the soil at the bottom of the hole. I was building on hard-packed gravel, which easily has a bearing capacity of more than 3,000 lb. per sq. ft. (psf).

The bearing capacity of a 10-in.-dia. tube in 3,000-psf soil is 1,650 psf ($0.55 \times 3,000$). The design load of each inside pier is 1,600 lb., so a 10-in. tube will work. However, by jumping up to a 12-in. tube, the bearing capacity becomes 2,370 psf ($0.79 \times 3,000$), which can carry the 1,600-lb. load more easily. For just a bit more concrete, I ensure the pier is well designed. I typically ignore the pier weight because there is enough fat in these calculations to justify this simplification.

The two outside-corner piers are required to bear only half the weight, but to simplify the work process I use the same size tubes for all four of the piers.

The depth you set the piers at depends a lot on the region of the country you're working in. In climates where frost is an issue, the minimum depth is established by code. For this project, the bottoms of the piers have to be 36 in. below finished grade.

Wherever you live, it is important to dig past soil that contains organic matter (topsoil) and any uncompacted fill. Organic matter decomposes over time and settles; loose fill also settles over time. In most cases, the depth of undisturbed soil is not known until the excavation is well under way.

Begin Layout with Deck Dimensions

Once I know the size and the number of piers I'm going to use, the next step is to lay them out on site. If the deck details

Everything You Need

A few tools, even fewer materials, and a little sweat will get most deck foundations out of the ground in less than a day.

Builder's tubes
80-lb. bags of ready-mix concrete
Garden hose
Foundation spikes
Batterboards
1/2-in. by 8-in. anchor bolts, nuts, and washers
Adjustable post bases
Post-hole digger
Digging bar
Electric concrete mixer
Stabila® plate level

aren't drawn on the plans, I sketch the outside deck framing to determine exactly where the center of the supporting posts are in relation to the outside dimensions of the deck. Then I use those locations to form a layout rectangle. I use batterboards and string to locate the exact center of the post, which is also the location for the anchor bolts that hold the post hardware in place.

Once the post locations are identified and marked with surveyor's paint, I remove the strings and dig the pier holes. When the holes are deep enough, I rough-cut the builder's tubes, drop them in, and replace the stringlines. I keep the tubes centered on the strings while they are backfilled, and I double-check the measurements with a tape measure.

Rather than try to cut tubes to exact height, I leave them long and pour concrete to the desired height inside the tube. In

How Many and What Size?

Three things affect the number and the size of piers you use: the way you frame the deck, the weight the deck is designed for, and the load-bearing capacity of the soil. For the deck I'm building, I chose to support the double rim joist with piers instead of a cantilevered approach that uses piers beneath a beam. I use the International Residential Code's design load for decks, which is 50 lb. per sq. ft. (40 psf live load, 10 psf dead load). Different soils have different bearing capacities (measured in psf); consult Table 401.4.1 of the IRC for the bearing capacities of different soil types.

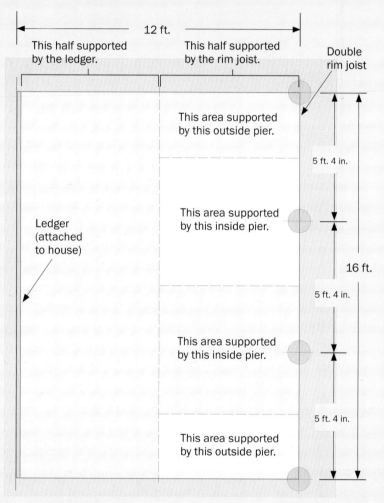

12 ft.

This half supported by the ledger.

This half supported by the rim joist.

Double rim joist

This area supported by this outside pier.

Ledger (attached to house)

This area supported by this inside pier.

5 ft. 4 in.

16 ft.

This area supported by this inside pier.

5 ft. 4 in.

This area supported by this outside pier.

5 ft. 4 in.

Step 1. Space piers evenly beneath the double rim joist

Because I'm using a double rim joist to support the floor joists, I support this 16-ft. deck with four piers.

Step 2. Distribute the deck's weight onto the piers

A 12-ft. by 16-ft. deck is 192 sq. ft. Multiply by 50 psf to determine the design load, 9,600 lb. Half of that weight (4,800 lb.) is carried by the ledger; the other half is carried by the piers. Because the corner piers carry only half the weight that the inside piers carry, dividing 4,800 lb. by three tells me the two inside piers must each bear 1,600 lb.

Step 3. Transfer the weight to the soil

For this project, I was working in hard-packed gravel, which I estimate to have a bearing capacity of 3,000 psf. Using the table here, I multiply the square-foot equivalent of each tube by 3,000 psf to find one that will work in this soil. A 10-in. tube will bear 1,650 psf, which is close, but I chose to bump up to 12-in. piers for peace of mind. To keep things simple, I made the corner piers the same size.

Tube dia.	8 in.	10 in.	12 in.	14 in.
Sq. ft.	0.35	0.55	0.79	1.1

most cases, I like the pour to come a couple of inches above the finished grade. If the piers are on a pitched elevation, the tops of the piers won't be level with each other. On this job, the finished grade was level, so I used a long level to carry the elevation across the piers.

After marking each pier with a small nail pushed through at the right height, I again remove the strings so that I can pour the concrete into the tubes. Once they're filled to the right height, I float the concrete smooth with a scrap of wood. Then I replace the string, and using a slight up-and-down motion to prevent air from becoming trapped, I insert the anchor bolts in their proper locations.

As a contributing editor for Fine Homebuilding *Rick Arnold has written nearly a dozen articles relating to framing. His most recent project involves installing four roof valleys on his new home.*

How Much Concrete Do I Need?

To pour the piers for an average-size deck, I use 80-lb. bags of concrete and an electric mixer, which rents for about $45 a day or sells for $250 or so.

For major pours, I have a concrete truck deliver a 2,500-lb. mix. Either way, the basic formulas shown here will help you to estimate the number of bags or cubic yards of concrete required based on pier size and depth.

Example
Size of tubes: 8 in.
Number of tubes: 8
Average depth per tube: 4 ft.
0.53 (8 × 4) = 17 bags

Tube size	Number of 80-lb. bags per foot	Cubic yards per foot
8 in.	0.53 bag	0.013 cu. yd.
10 in.	0.8 bag	0.02 cu. yd.
12 in.	1.2 bags	0.03 cu. yd.
14 in.	1.6 bags	0.04 cu. yd.

Sources

Rousseau Co.
1718 13th St.
Clarkston, WA 99403
800-635-3416
www.rousseauco.com
Reusable batterboard system
Cost for a set of four: $64

Stabila, Inc.
332 Industrial Dr.
South Elgin, IL 60177
800-869-7460
www.stabila.com
Plate level

FOOTINGS
Bigfoot Systems®
RR 2
Mahone Bay,
Nova Scotia
Canada
B0J 2E0
800-934-0393
www.bigfootsystems.com
Available from 20 in. to 36 in. dia.
Cost: $18 to $75 per form

The Footing Tube
28 Amberwood Lane
Fredericton, NB
Canada E3C 1L7
506-452-8919
www.foottube.com
A builder's tube and spread footing in one. Top diameter sizes range from 6 in. to 12 in.
Cost: $33 to $47 per form

Redibase
106 Papineau Rd.
Fulford, QC
Canada J0E 1S0
450-242-5805
www.redibase-form.com
Available in 24 in. dia.
Cost: $9 to $12 per form

Square Foot
37 Talcott Rd.
Williston, VT 05495
877-924-2323
www.sqfoot.com
Available from 22 in. to 32 in. dia.
Cost: $16 to $30 per form

Use Two Lines for a Dead-On Layout

With the ledger location transferred to grade level, I can measure out from the house foundation and run a stringline to represent the centerpoint of the piers. A single line parallel to the house intersecting a line perpendicular to the house locates the center of the far-corner pier. Measurements for the rest of the piers are taken from this intersecting point. Batterboards help to set the lines accurately (see "Tool Tip," on the facing page).

Plumb down from a high ledger. With a Stabila plate level, I carry one end of the ledger down to the grade. I drive a stake into the ground here to anchor a line that will run perpendicular to the house.

The pier centerline runs parallel to the house. The batterboards I use make it easy to adjust the string until it's exactly the right distance from the house foundation. I set the batterboards a couple of feet beyond the corner-pier locations so that the boards won't be disturbed when holes are dug.

$A^2 + B^2 = C^2$. Pulled diagonally from the foundation, my tape forms the hypotenuse of a right triangle. A helper shifts the line that extends from the house to intersect with the right measurement, identifying the center of the far-corner pier.

Mark piers with paint. Measure the remaining piers from the far-corner pier. A dot marks the centerpoint and a rough circle highlights where to dig. Pull the stringlines and prepare to dig, but keep the batterboards in place.

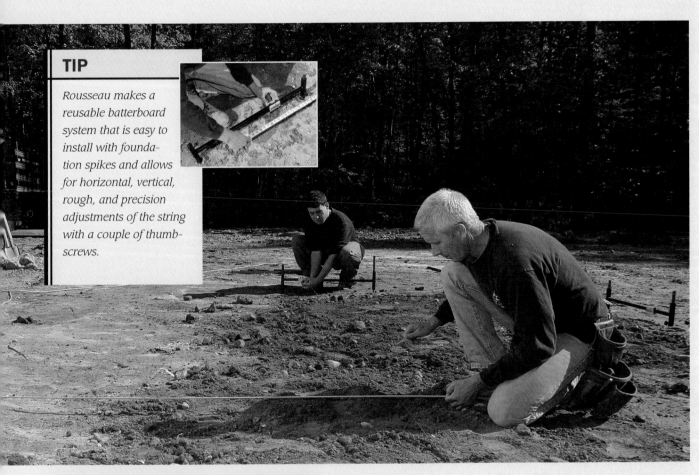

TIP

Rousseau makes a reusable batterboard system that is easy to install with foundation spikes and allows for horizontal, vertical, rough, and precision adjustments of the string with a couple of thumb-screws.

5 **The best holes have no rocks.** But just in case you encounter a few, make sure to have a long digging bar in addition to a post-hole digger. Take care not to disturb the batterboards or their settings, because you'll have to reattach the strings later. Dig down deep enough that the bottom of the pier rests on undisturbed soil below the frost line.

TIP

If excavating multiple holes, consider renting a gas-powered auger. A one-person auger, shown here, costs about $50 a day. The two-person version rents for about $70.

Fine-Tune the Layout Before and After the Pour

Once the holes are dug, put the stringlines back on the batter-boards. When setting each builder's tube, use the lines and a tape measure to center them according to layout, adjusting the hole locations as needed. Take the time to check the tube locations often as you backfill to keep them on layout. After all the fill is in place and the final layout check is made, fill the tubes with concrete and insert the anchor bolts.

Double-check the corners. I spend a little extra time checking the location of the final corner pier to make sure that it's in the right spot, because I won't get a chance to move it once the concrete is poured. Use a nail to mark the finished height of the piers, keeping it a couple of inches above the finished grade. If you need to have piers all at the same height, use a long level or a transit to locate their finished height.

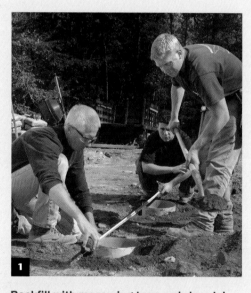

Backfill with measuring tape and shovel. I cut the builder's tubes so that they stick out a few inches above grade when placed in the hole. To make sure a tube is placed precisely, I hold it on its layout while a helper backfills. Pack the soil around the tube every so often as you go.

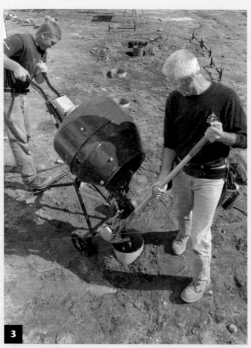

3

A shovel makes up for bad aim. Fill the tube with concrete until it reaches the nail. The concrete should be just slightly on the wet side, about the consistency of thick oatmeal. As the concrete is poured into the tube, a helper uses a shovel to agitate the mix every 8 in. to 10 in. to work out air pockets.

> **TIP**
>
> *A mixer does the most difficult work of mixing the concrete. Just dump in the mix, turn it on, then add water.*

4

Place anchor bolts accurately. Once all the piers are poured, I go back and insert anchor bolts in the center of the piers. I measure from the line running perpendicular to the house to set anchor bolts accurately. Be sure to leave the threads high enough so that a post base, washer, and nut can be added later.

5

Adjustable post bases allow for final tweaks. After the concrete is cured completely, I attach adjustable post bases. I like to use Simpson® ABA-style bases because they allow me to fine-tune the post location after the post is attached.

Soil: The Other Half of the Foundation

■ BY ROBERT M. FELTON

Most builders and architects are familiar with the problem of building settlement: the chimney that won't stop going down, the wall cracks that keep opening up, the older home that suddenly begins to exhibit movement for no apparent reason. Everyone in the building industry has a story about a fix that "shoulda done it," but didn't.

Preventing settlement problems begins with the recognition that the soil a foundation rests on is part of the foundation system; it's a building material, just like the 2×4 studs that frame the house. The fact that you can't go to a lumberyard and select this building material—that in most cases you're simply going to use whatever you happen to have—makes it especially important that you recognize differences among soil types, know something of the way soils respond to building loads, and be able to identify potential problems.

Differential Settlement Is the Real Enemy

A few things need to be understood about settlement. First, all houses settle. The amount may be so small as to be undetectable or may be so uniform as to leave no signs, but it unquestionably happens. Second, because of natural and construction-related variations in soil properties, not every point on a foundation settles the same amount.

To avoid problems with entrances and utility connections, total settlement must be minimized. To avoid racking door frames and cracking walls, you must prevent differential settlement, the difference in settlement between various points on the foundation. The distinction between total and differential settlement is important. The Palace of Fine Arts in Mexico City, for instance, has settled several meters without significant distress to the structure and remains in service because the settlement has been uniform. The Leaning Tower of Pisa, on the other hand, is useless for anything but the Kodak® moments of tourists.

Finding out what's down there. Taking samples from a series of drillings enables engineers to determine subsurface soil characteristics.

Failure to compact is what leads to cracking sidewalks and uneven driveways.

Elementary Research Can Dig Up Most Problems

Fortunately for homebuilders, the loads involved in most residential construction are relatively light. Following time-proven procedures and steering clear of some common misconceptions will keep you out of trouble in most cases.

Two Ways to Look at the Same Piece of Land

Soil maps from the U.S. Department of Agriculture (see the map at bottom left) are actually aerial photographs with soil types superimposed on them. They can alert builders to problem soil conditions before they start digging. Topographic maps from the U.S. Geological Survey (see the map at bottom right) can reveal wetlands or intermittent streams. Serving as snapshots in time, both maps can provide valuable insight into human activity that has taken place on land.

USDA Soil Map

USGS Topographic Map

City dump?
The soil descriptions that accompany this map indicate that except for the suspicious area in the center ("Ma" signifies man-made soil), all the soils delineated in this close-up are builder-friendly sand. The two straight, bold lines to the left of the dump represent dammed lakes.

What city dump?
Although a topographic map is generally the better indicator of previous development, a builder who relied on only a topo map in this case would not be able to tell that the lakes were man-made and, more important, would have no idea that his dream land was once the city dump.

For starters, you can learn a lot about soil conditions on your site by taking advantage of public-sector resources. The United States Department of Agriculture (USDA) has prepared soil maps for most of the country. Available at no cost at any local USDA branch office, these maps superimpose soil-type delineations over aerial photographs (see the drawing above left). By studying these maps and the soil descriptions that accompany them, you can find out information such as whether your site might have a high groundwater table or whether problematic soils—such as shrink/swell-susceptible clays—might lurk beneath the surface.

Having been taken 30 years ago or more, aerial photos often reveal evidence of unsuspected development or manipulation of the site. An even better source for this type of information is a topographic map from the United States Geological Survey (see the drawing at right on p. 56). This map may reveal abandoned cemeteries, farm ponds, wells, or the long-forgotten town dump. A topographic map may be purchased for a few dollars at outdoor-sporting-goods stores or downloaded at no cost (www.topozone.com).

Don't forget to check with the local building and engineering departments as well; they often have a wealth of local information and experience, which they are usually happy to share. Developers who have built close by or homeowners on adjoining lots are other good sources of information.

Virgin Soil Is Not Always Virtuous

If your research unearths potential problems, that's the time to bite the bullet and consult an expert. If you uncover no history of activity that may have left problems behind, that's probably good news, but there may still be things that need attention. "Virgin soil" isn't inherently problem-free. Muck (decomposed organic material), for instance, may have been undisturbed since Mother Nature deposited it, but building on it is always a bad idea.

The Other Half of the Foundation

To prevent settlement problems, soil that is stable and compacted must extend at least half the distance from the base of the footing to the influence depth (the farthest distance beneath the footing that its weight is felt by soil particles). Strip foundations (such as the wall on the right) stress the soil more than square foundations (such as the pedestal on the left) and thus have greater influence depths.

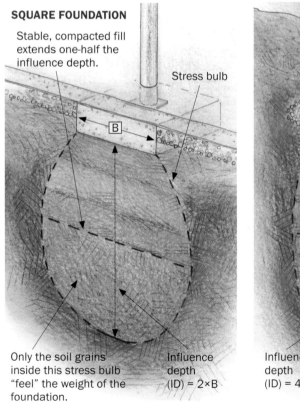

SQUARE FOUNDATION

Stable, compacted fill extends one-half the influence depth.

Stress bulb

Only the soil grains inside this stress bulb "feel" the weight of the foundation.

Influence depth (ID) = 2×B

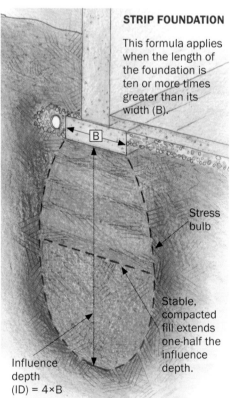

STRIP FOUNDATION

This formula applies when the length of the foundation is ten or more times greater than its width (B).

Stress bulb

Stable, compacted fill extends one-half the influence depth.

Influence depth (ID) = 4×B

Clay can also be troublesome. The strength of clay soils varies inversely with changes in moisture content: The greater the moisture, the weaker the soil. If clay materials underlie your site, the site plan must provide for positive drainage that will direct surface water away from the structure and paved areas; otherwise, water may penetrate and weaken the supporting soil. This is, in fact, a common cause of post-construction settlement problems. If site constraints make it impossible to direct runoff away from the driveway, you should plan to provide lateral drains alongside the driveway to prevent water from accumulating beneath the pavement. Foundation drains must also be carefully designed to carry groundwater well away from the structure. These measures aren't cheap, but they cost less than repairs, ill will, and neighborhood gossip.

Bridging the Gaps

What if there's a wet, loose, soft, low, or mucky spot on the site? Can you bridge it with fill? Geotechnical engineers like me have a couple of easy rules of thumb that we

When Should You Call in a Soils Engineer?

Many builders and architects are reluctant to hire engineers to perform subsurface investigations, and that's easy to understand: The cost of a house can go up a few thousand dollars. Despite the cost, there are circumstances where consulting a soils engineer is a wise investment.

• If you're wondering how on earth (to belabor the point) you're supposed to build a house on a lot with a steep slope, then you need an engineer to determine the soil properties and to evaluate the stability of the slope.

• Evidence of previous earthwork at the site, especially filling, requires careful investigation. The site may host buried organic materials (i.e., muck), debris such as demolition rubble or abandoned vehicles, or simply dumped fill. Any of these things can cause severe settlement problems.

• A local history of the presence of clay soils that are susceptible to shrinking and swelling with changes in moisture content also requires careful investigation and, usually, specialized design services and use of an out-of-the-ordinary foundation.

• Encountering groundwater or weak soil while excavating for foundations may indicate a potential for settlement or instability in the future. The problem area should be carefully delineated by an expert and remedied after close consultation.

Consult your experience: If you're crossing your fingers and hoping the soil conditions won't cause a problem, you're probably right to be uneasy. Call in a specialist. To locate an engineer in your area, contact the American Society of Civil Engineers.

Not exactly wine tasting. Soils engineers inspect core samples they've taken using the split-barrel sampler on the left.

refer to when called to a site and asked what to do. After that, it gets complicated.

Rule one: Never fill a low spot with surface strippings, shrubs, or woody debris, no matter how much clean fill will be placed on top of it. It might take years, but the organic material will inevitably decompose and cause settlement.

Rule two: Muck must be removed. Although it's possible to force sand and gravel fill into the muck and create a stable mixture, that's not a reliable solution. What usually happens is that pockets of weak material become scattered throughout the reworked material, and these areas slowly compress over the years.

To control settlement problems, soils engineers want the foundation to rest on stable, compacted material that extends at least half of the influence depth, the distance beneath the footing that its weight is still "felt" by soil particles (see the drawings on p. 57).

Square foundations (such as those beneath a Lally column) stress the soil to a depth equal to about two times the foundation width. A 2-ft. by 2-ft. pedestal, for example, is "felt" by soil particles to a depth of about 4 ft. below the foundation.

A strip foundation, one whose length is ten or more times greater than the width (e.g., a wall foundation) stresses the soil to a depth equal to about four times the foundation width. Thus, an 18-in.-wide foundation is "felt" by soil particles as much as 6 ft. below the foundation. The maximum stress for either type of foundation occurs at a depth of about one-quarter of the influence depth.

What all this means to the builder is that for the pedestal I just described you must provide at least 2 ft. of competent support; for the strip, you must provide at least 3 ft. If that can't be done with the existing soil, you should plan to remove and replace the undesirable material and restore the site to grade with engineered fill.

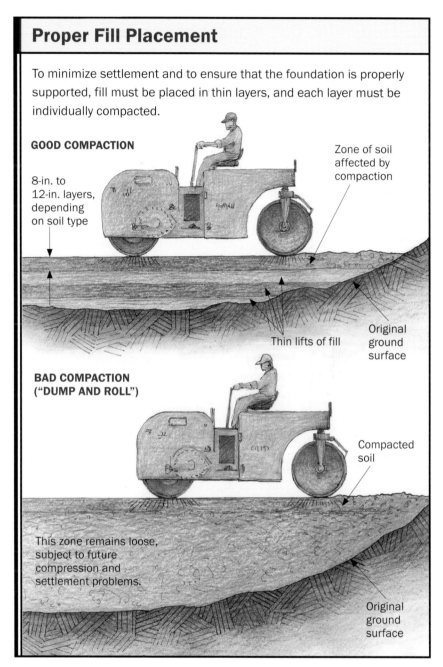

Proper Fill Placement

To minimize settlement and to ensure that the foundation is properly supported, fill must be placed in thin layers, and each layer must be individually compacted.

GOOD COMPACTION

8-in. to 12-in. layers, depending on soil type

Zone of soil affected by compaction

Thin lifts of fill

Original ground surface

BAD COMPACTION ("DUMP AND ROLL")

Compacted soil

This zone remains loose, subject to future compression and settlement problems.

Original ground surface

Self-Compacting Soil Has Not Been Invented Yet

One of the misconceptions I often encounter is the belief that soil will densify and strengthen if it's merely dumped on a site and left undisturbed for several months—that fill dumped in a low area in the fall

Compacting Soil against Foundation Walls Takes Care

To avoid damage from the horizontal loads imposed by compaction, heavy equipment should be kept away from the wall a distance of at least two-thirds the unbalanced height of the fill.

10-in. concrete wall

Minimum two-thirds H

Compact in this area using light hand-operated equipment.

H

1 ft.

Concrete footing   4-in. concrete slab

will be ready to support construction in the spring without any compaction. To this opinion, I always respond, "'Fraid not." To understand why, think back to your high-school physics class: An object at rest tends to remain at rest unless acted on by an outside force. Loosely dumped soil does not densify and strengthen by itself.

It is particularly important that fill be placed in thin layers so that the densification effect of the compaction equipment is felt all the way to the bottom of each layer (see the drawing on p. 59). The maximum thickness for each layer varies depending on soil type: ordinarily about 12 in. for sand and 6 in. to 8 in. for clay. The required degree of densification is usually set forth by the local building code and specified as some percentage of maximum dry density as determined by one of several standard methods.

The moisture content of the fill material must also be controlled. If the moisture is too low, it is difficult for individual soil grains to realign themselves into the densest configuration; adding moisture lubricates the grains and makes realignment easier. But if there is too much moisture, the soil becomes unstable under the influence of compaction equipment because a portion of the compactive effort will be borne by the water between the soil grains and result in a water-bed-like rolling of the soil. Granular soils such as sand are most desirable for use as fill because their moisture content can be easily tweaked.

Keep Heavy Equipment away from Foundation Walls

Compacting soil against below-grade walls also requires special care, lest the horizontal load imposed by the compaction equipment

Hairline crack or structural disaster? A crack monitor is used to track crack movement over a period of time. The pattern of movement enables an engineer to determine when and what remedial action may be necessary.

damage the walls. A good rule of thumb is to keep heavy equipment away from the wall a distance of at least two-thirds the unbalanced height of the fill (see the drawing on the facing page). In other words, if the fill on the outside of the wall is 6 ft. higher than the fill on the inside, the big rollers should be kept at least 4 ft. away from the wall.

The strip of ground adjoining the wall should be compacted using small, hand-operated equipment. But make no mistake: It should be compacted; failure to compact leads to cracking sidewalks and uneven driveways. It is equally important to assure that fill placed in utility-line excavations be properly placed and compacted. Improper placement of fill by the plumbing contractor, for instance, not only causes drain problems but frequently leads to exterior-wall settlement and cracking over buried drain lines.

Troubleshooting

In spite of your best efforts, what if the new homeowners call your office and demand that you come over right away to inspect a crack they've just noticed? You not only have a public-relations problem, because previously unnoticed hairline cracks suddenly become subjects of concern to the buyer, but also a real technical problem: You've got to figure out what has happened.

Most cracks are minor and insignificant, a consequence of the settlement and shifting that all houses undergo. In these cases, you need to explain a few things to the homeowners. Ideally, you would have prepared them for these possibilities beforehand:
• The act of building the house changed the local groundwater conditions. In particular, the shallow soil zone beneath the house is drying. The minor settlement that results can be enough to cause hairline cracking.
• When you painted the house, it was empty. When the owner took possession and brought in the home gym, water bed, and baby grand, this stress inevitably caused

flexing of virtually all the structural members. This, too, can cause hairline cracking.

Of course, there may be times when simple explanations don't suffice, when something strange seems to be happening. The best way to identify the cause of the problem is to install a crack monitor (Avongard Products), a two-piece, specialized ruler that is semipermanently mounted over a crack. Using a crack monitor spares you the work of trying to decide by eye and memory whether a crack has grown or changed. If it appears you've got a real problem, the data will be invaluable to whoever is trying to figure out what's causing the crack and what ought to be done, because different settlement mechanisms leave distinctive signatures, which become apparent when the data are graphed.

The correction of foundation problems requires thorough investigation by experts. You should not hire a grouting contractor to "mud-jack" (pump concrete beneath a distressed area) until the cause of the distress has been positively identified. If a house corner is settling due to the presence of buried, compressible organic material, for example, pumping a yard or two of concrete into the soil immediately beneath the foundation will increase the load over the soft material and thus increase settlement. The wall may look fine when the contractor cleans up the job site and leaves, but more irate-homeowner phone calls are inevitable.

When making repairs, just as when beginning a project, you should be guided by a single rule: The best defense against future problems is doing it right the first time.

Robert M. Felton is a consulting geotechnical engineer and freelance writer in Wake Forest, North Carolina. He may be contacted via email at bobfelton@gmail.com.

Start with Batterboards

■ BY JIM BLODGETT

You're finally ready to build that new deck, porch, patio, or addition. With a permit in your pocket, money in the bank, and a shovel in your hand, it hits you: Now what? How do you make the transition from concept to concrete? Batterboards. Batterboards are like bookmarks. They're placeholders for strings that describe the dimensions of a structure. You can use batterboards as a job progresses to reduce compounding error.

Build with Screws, Brace with Stakes, and Walk Away

To keep batterboards accurate, place them safely away from construction traffic, and build them sturdily. Use braces. I usually place batterboards within a couple of feet of the foundation and discourage the flow of foot traffic with strategically placed piles of demolition debris, dirt, or lumber.

I use two types of batterboards for most small projects: a three-legged freestanding type for outside corners, and a single 2x4 screwed horizontally (and level) to the side

Step 1: Choose your triangle from the table (below) of multiples. Select the triangle nearest to your project's footprint; bigger is better.

Foundation mark
18-ft. mark

SHORT LEG

HYPOTENUSE

LONG LEG

SHORT LEG	LONG LEG	HYPOTENUSE
3	4	5
6	8	10
9	12	15
12	16	20
15	20	25
18	24	30
21	28	35
24	32	40
27	36	45
30	40	50

Five Steps to Squaring Success

One goal of batterboards is to define a footprint with square corners. Most people use the Pythagorean theorem ($a^2 + b^2 = c^2$) to check that the corners are square. They do so by measuring 3 ft. along one side, 4 ft. along the other, and checking for a 5-ft. diagonal (or hypotenuse) between those two points. But those dimensions (3-4-5) are too small to square up a larger corner accurately. It's better to use multiples of 3-4-5 (see table) that define a right triangle that's close to your building's dimensions.

The author prefers nylon string over cotton because he can pull it tighter and also because it better resists fraying and breaking.

2×4 batterboard screwed to house

Step 3: Define the long leg of the triangle.
Often a door, a window, or a wall dictates where a project begins. If not, pick a spot and call it point A. Measure 24 ft. to point B. Screw 2×4 batterboards to the house at points A and B.

Strings normally define the outside of a foundation, but they can also help to locate deck-support piers.

Set string in sawkerf.

Step 4: Measure the hypotenuse.
Pull a tape measure from point B, and find where the 30-ft. mark intersects the reference string (mark this point with ink). Now run a string from point A through this mark and fasten the string to the batterboard.

Step 5: Define the project's footprint. From points A and C, measure and mark the width (points D and E). From points D and E, measure back and mark the depth.

Foundation mark

Batterboard

Brace with diagonal boards.

Step 2: Set a reference string parallel to the house at a distance equal to the triangle's short leg (18 ft.).

Screw, don't nail, connections.

Drive the upright at least half its height into the ground.

Set horizontal boards level to maintain level strings.

Double-edged point guides the upright straighter than a single-edged point.

Quick-Release Knot Holds String Tight

Loop string around one finger, twisting that finger several times as you pull your hands apart.

Place the loop over a nail, pulling on the free end of the line while feeding the fixed end toward the nail.

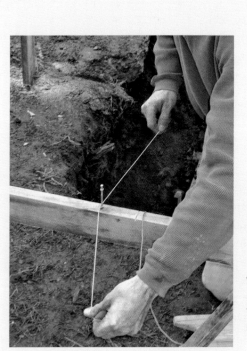

When the string is banjo-tight, pull the free end back toward the nail. The twists will ball up into a knot. To loosen, pull the free end in the opposite direction.

of the existing house. Rather than nails, I use screws so that I don't loosen the braces on freestanding batterboards, which lessens accuracy. Deck screws are great for this, but drywall screws are fine.

Use a Builder's Level Once, Then Leave It in Its Case

Installing batterboards that are level with each other allows me to use a level string to set the grade on excavations, footings, and foundations. I generally use an elevation higher than where the top of the foundation will be. I level batterboards with a water level or a builder's level.

After the building lines and levels are determined and double-checked (and verified and triple-checked), I use a handsaw to score just deep enough through the top edge of the batterboard to hold the string. No matter how many marks, arrows, offsets, or elevation notes are written on the batterboards (and over the course of a job, there can be a lot), I can find the building lines quickly.

I take all measurements from these saw kerfs, and I mark them on the batterboards. In addition to the inside and outside faces of the footing and foundation walls, I lay out offsets from the building lines so that I can set up a parallel string quickly. These offset lines allow me to climb in and out of the excavation without doing the limbo to avoid lines all day.

Jim Blodgett has written for Fine Homebuilding *about lifting heavy walls, working with MDF, retrofitting skylights, and trimming windows.*

Pouring Concrete Slabs

■ BY CARL HAGSTROM

Liquid stone. It's an image you might think describes placing concrete, and to some extent, it does. But there's more to it than backing up the ready-mix truck, opening the spigot, and letting the concrete flow out until the forms are full.

The applications of concrete are almost limitless, but here I'll focus on residential slabs. About half of the new homes currently built in the United States start with full-basement foundations, and virtually all of these basements have concrete floors. For the most part, these floors consist of 4 in. of concrete placed over 4 in. or more of crushed stone. Concrete floors in garages are similar, except they are sometimes reinforced with wire mesh or steel. But whether it's a basement or a garage slab, the way you place the concrete is the same.

Ordering Concrete

Concrete is sold by the cubic yard, and calculating the amount you need is simple: length times width times depth (in feet) divided by 27 equals cubic yards. Most concrete trucks max out at 9 yd., and if your floor will require more than 9 yd. (the average floor uses about 19 yd.), tell your supplier to allow about an hour to an hour and a half per truckload so that all the trucks don't arrive at once.

But a word of caution. Running out of concrete is like running out of champagne at a wedding: If you can't get more real soon, you're headed for trouble so don't be stingy with your concrete estimate. You're a lot better off with half a yard left over than a quarter yard short.

Once you've told your concrete supplier how much concrete, you'll have to tell them what kind. Concrete is made up of four basic ingredients: cement, sand, stone, and water. Depending on the proportions of the ingredients, the strength can vary considerably. Compressive strength, measured in

Using a screed rail. Two workers use a magnesium straightedge, or screed rail, to level freshly placed concrete. The ends of the rail glide over previously leveled concrete strips, called wet screeds. A third worker rakes the concrete behind the screed rail to adjust for high and low spots.

pounds per square inch (psi), is the method used to evaluate the performance of a given mix. Generally, the higher the cement content, the higher the compressive strength. Most residential concrete has a compressive strength from 2,000 psi to 3,500 psi.

You'll also need to specify the slump, or wetness, of the mix. A slump of 4 to 5 is about right for slabs, whereas a slump of 2 to 3 is normal for piers, which don't need to be worked, so the concrete can be stiffer.

Placing the Slab

Arrive early on the day of the pour and use a water level or a transit to snap chalklines on the foundation wall at finish-floor height (usually 4 in. higher than the stone). The lines help you level the concrete along the walls.

You should also lay out the vapor barrier at this time. Six-mil polyethylene works well, but if you're concerned about punctures from traffic during the pour, a puncture-resistant, cross-laminated product is available, called Tu-Tuf® from Sto-Cote Products, Inc.

If you elect to use wire-mesh reinforcement, this is also the time to lay it out. Wire mesh doesn't prevent cracking, but it will help keep hairline cracks tight, even as the temperature varies. Typically, a basement slab isn't subjected to wide temperature swings. Therefore, a basement slab placed over a properly prepared stone base doesn't require wire mesh. Garage slabs, on the other hand, typically experience harsher weather conditions, and wire mesh may be used as temperature reinforcement. But wire mesh won't be effective unless it's placed midway in the thickness of the slab, so be sure to use wire high chairs, which hold the reinforcement up off the stone during the pour.

The ready-mix truck arrives, and the driver asks, "How wet do you want it, Mac?" Drivers routinely ask about adding water to soften the mix. When a mix is too stiff, it's physically difficult to work and presents problems when it's time to float and finish the slab. To get a smooth, hard, dense finish on top of the slab, the mix has to be workable. As mentioned earlier, however, the wetness was determined when you specified the slump of the mix. And as any structural engineer will tell you, when you add water to concrete, you lower the final strength. The issue of water content in concrete is critical; many concrete companies require that you "sign off" on the delivery slip when requesting additional water so that they have a record of your compromising the rated strength of the mix.

If the first few wheelbarrows of concrete are difficult to work, have the driver add water to it—in small amounts. You can always soften the mix by adding water, but you can never dry it if it becomes too wet.

Leveling with Wet Screeds

There are many ways to place a basement slab. If you've never placed one, ask some masons about the techniques they use. If you have placed a few slabs, don't be afraid to try a different method; you may discover a system that you're more comfortable with. But whatever approach you take, follow a logical progression: Don't trap yourself in a corner. I prefer to use wet screeds as guides to level the slab (see the photo on p. 67).

Wet screeds are wet strips of concrete that are leveled off at finish-floor height and used to guide a straightedge, or screed rail, as you level the slab. If you've ever watched a sidewalk being placed, you've seen concrete placed between two wood forms, a screed rail placed on top of those forms and sawed back and forth to strike the wet concrete down to the level of the forms. Wet screeds guide the screed rail in places where there are no wood forms, such as against an existing concrete wall or in the middle of a slab.

Where you start with your wet screeds depends on the layout of the slab. In a typical rectangular basement with the walls already in place, a wet screed is placed around the perimeter of the foundation, and a second wet screed is placed down the center of the foundation (see the top photo on p. 71), parallel to the longer dimension of the foundation. On a bigger slab you might need more wet screeds; the determining factor is the length of the screed rail you'll be using.

Placing the wet screeds around the perimeter of the foundation is simple. Use the chalkline you snapped at finish-floor height as a guide to level the concrete at the wall (see the bottom photo at right). As the concrete is placed, either from a wheelbarrow or directly from the chute, use a magnesium hand float to push and level the concrete to the line. Be sure you don't cover up your chalkline as you place the concrete. Dump it near the wall and bring it up to the line with the float (see the top photo at right).

Establishing the level of the center screed requires that you drive pins about 8 ft. apart at the level of the finish floor; 16-in. lengths of ½-in. rebar work well. Try to set these pins immediately before the pour, using a transit or a stringline, and cover them with upturned buckets so that no one trips on the pins. Place and level a pad of concrete around each pin, then fill in the area between the pads with concrete and use a screed rail, guided by the pads, to level the area between them. As you complete each portion of this center screed, drive the pins a few inches below the surface with a hammer and fill the resulting holes with a little concrete.

Raking and Striking

To fill in the areas between screeds, place and rake the concrete as close as possible to finish level before striking with a screed rail. Placing too much material makes it difficult to pull the excess concrete with the screed rail as you strike off, and the weight of the excess concrete can distort a wooden screed rail. If you starve the area between the

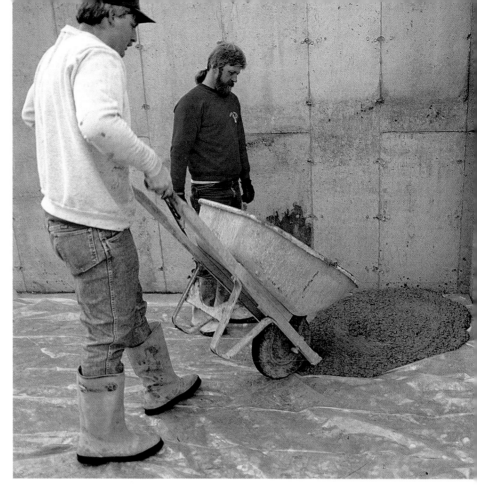

First mud. With the vapor barrier in place and the chalklines snapped, the first load of concrete is dumped in the far corner of the foundation. The mason dumps the concrete away from the wall so that he won't cover the chalkline.

Establishing a perimeter screed. A magnesium hand float is used to push the concrete up to the chalkline. This strip of wet concrete, placed along the foundation walls, is a perimeter screed.

Be Prepared for the Pour

Be prepared for the day the concrete is scheduled to arrive by first surveying your situation.

Do you have a grade-level door, or will you need to chute the concrete through a basement window? Will the ready-mix truck be able to get next to the house, and if not, will the manpower be available to transport the concrete in wheelbarrows? Pushing one wheelbarrow full of concrete uphill is possible for some, but making 30 trips uphill is a job for the John Henry type. It never hurts to have more help than you might need because concrete is always a rugged day's work.

A little rain the night before can turn a dry approach into a muddy nightmare. I call my supplier several days in advance and say that I'm shooting for next Thursday, for example, but that I'll call first thing Thursday to confirm. If conditions are terrible, I reschedule.

Remember, concrete waits for no one. From the minute it leaves the plant, it has a finite time before it sets up, and just about any builder can come up with a horror story describing a pour that got away.

Teamwork is the name of the game. Establish each person's role well ahead of the pour, and do your best to stick to the plan.

screeds, you'll constantly be backtracking through freshly placed concrete, filling in low spots and rescreeding.

Using a screed rail, strike off the concrete with the perimeter screed and the center screed as guides. Your path of escape will determine the placement and the size of your screeds, but generally speaking, you progress in about 10-ft. or 12-ft. sections of slab.

The person raking the concrete can make or break the pour. As the wheelbarrows are dumped, the raker should nudge the concrete to the plane of the finish floor, eyeing the placed concrete like a golfer lining up a putt, and noting any mounds or valleys that will create problems as the screed rail works across. As the concrete is struck off, an alert rake person will pull away any excess concrete accumulating ahead of the screed rail (see the top photo on the facing page) and push concrete into any low spots.

At this stage of the pour, with five or more people working, teamwork is the name of the game. Establish each person's role well ahead of the pour, and do your best to stick to the plan.

Striking off the concrete with the screed rail is the last step in placing the concrete and the first step in finishing it. A good, straight 2x4 will work well, but magnesium screed rails, available in various lengths, will perform better. No matter which you choose, working a straightedge back and forth is a lot like running a two-man saw. The work is done on the pull stroke, and you have to be aware of your partner's progress. Wear rubber boots because, standing alongside the wet screeds, you'll often be wading through concrete. (For anyone tempted by the prospect of a barefoot frolic in concrete, be warned that concrete is caustic and will corrode your skin.)

As you saw the screed rail back and forth, let it float on top of the wet screeds, keeping an eye open for low spots and stopping when excess concrete dams up ahead of the screed rail so that the raker can pull off the excess.

Bull Floating

As the pour progresses, it's necessary to smooth the surface of the leveled concrete with a magnesium bull float (see the bottom photo on the facing page). The length of the float's handle determines the size of the area you can smooth effectively. For example, a bull float with an 18-ft. handle will easily float a 10-ft. or 12-ft. section of a slab. Bull floating levels the ridges created by the screed rail, but more important, it brings cement and sand to the surface of the slab and pushes stones lower.

Water is the lightest ingredient in concrete and quickly finds its way to the surface as you jostle the mix around with a bull float. As the water rises to the surface, it also brings some cement and sand with it.

Raker's role. As the screed rail levels concrete between the perimeter screed (right) and the center screed (left), the raker pulls away excess concrete or fills low spots. A rebar spike set at finish-floor height and subsequently driven below the concrete's surface establishes the center screed's level.

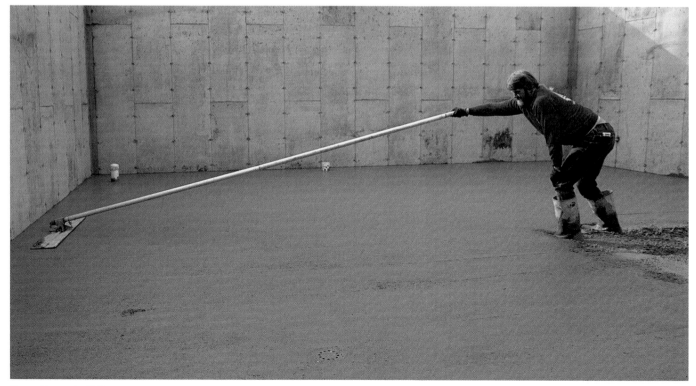

Operating a bull float. After the concrete is screeded, a bull float pushes stones down and brings up the fines—sand and cement—that make a smooth, finished surface. To use a bull float, lower its handle as you push it away, and lift the handle as you pull it back.

These are the fines (sometimes called fat or cream) that provide a stone-free medium for troweling to a smooth finish.

Although its size is intimidating, a bull float works about the same as a hand trowel. The trick is to keep the leading edge of the bull float inclined above the surface of the slab by lowering the bull float's handle as you push it away and raising the handle as you pull it back. Some masons jiggle the handle as they move it out and back to jostle more fines to the surface. The ease of final troweling depends on how well the slab has been bull floated.

Hurry Up and Wait

Once the slab is placed and bull floated, it's time to sit and wait. The first stage will be the evaporation of the bleed water, water that rises to the surface as the slab sets up. Depending on the weather conditions and the consistency of the mix, this time can vary from one hour on a hot, dry day to 10 hours on a cool, damp day.

But keep in mind that when concrete starts to set, it waits for no one. There is a small window of opportunity in which you can work the slab, and if you happen to be out for coffee when the concrete starts to set up, you'll learn an expensive lesson. Unless you're a veteran finisher, don't ever leave the pour; you may return to a problem whose only solution is a jackhammer.

Floating from Kneeboards

Once the bleed water has evaporated, work the slab. Some slabs (in crawlspaces, for example) are acceptable with just a coarse, bull-floated finish. But these finishes tend to dust over time; that is, concrete particles come loose from the coarse slab surface whenever it's swept. Additional finishing compacts the surface so that the slab won't dust.

You may have seen professionals using a power trowel to float and finish larger slabs. A power trowel works like a lawn mower without wheels. It rides on rotating blades that smooth the surface of the concrete. If you're inexperienced, however, or if the slab is small, you're better off finishing it by hand. And even professionals still use hand floats and trowels at the edges of the slab and around projections because a power trowel will only finish to within a few inches of these spots.

Hand finishing is commonly done from kneeboards, which are like snow shoes for the still-wet slab (see the left photo on p. 74). They let you move around on the slab without sinking. To make a simple pair of kneeboards, cut two pieces of ¾-in. plywood 2 ft. square, and tack a 2x2 strip at one edge of each piece.

It's difficult to describe just when the slab is ready for hand floating, but it may help to think of the slab as drying from the bottom up. If you set a kneeboard on the slab, and it sinks ¾ in. when you step on it, you're too early; if it fails to leave a mark, you're too late.

As soon as you can easily smooth over the tracks the kneeboards leave behind, the slab is ready for the first hand floating.

Test the concrete at the area where the pour started because it tends to be ready first. If any areas of the slab are in direct sunlight, you can bet they'll be ready long before the shaded areas are. At any rate,

Tips for Pouring in the Weather

Temperamental is a literal description of concrete. Temperature, along with humidity, influences the pour more than any other factor.

Hot-Weather Pours

When it's hot, and the humidity is low, every minute is important. If you spend time fussing around, when the last wheelbarrow of concrete is finally off the truck, the first section of floor you placed will probably be hard enough to walk on.

Here are some strategies that help in hot weather:

- Even if a polyethylene vapor barrier is not required, use one. It blocks the moisture from dropping through the subgravel.
- Have lots of help available. The sooner you get the truck unloaded and the concrete leveled, the better your chances will be of getting a good finish.
- Have two finishers working the slab: one with a magnesium float, and another following behind with a steel trowel.
- Although it compromises compressive strength, consider using a wetter mix to buy a little more working time.
- If more than one truckload is needed, coordinate the arrival times carefully. If a fresh truckload of concrete has to sit and wait an hour while you finish unloading the first truck, you may find that concrete from the second truckload will set up before you're ready for it.
- Areas that receive direct sunlight set up much quicker than shaded areas.
- Start wetting down the slab as soon as the final finish has set. Few things will weaken concrete as much as a "flash" set, where the concrete dries too quickly.

Cool-Weather Pours

When the temperature is cool, concrete initially reacts in slow motion. After the slab is placed, and the bleed water slowly evaporates, you'll wait hours for the slab to tighten up enough to start hand troweling. When it's finally ready to be troweled, you'd better be there because that window of opportunity for finishing doesn't stay open much longer on a cool day than it does on a warm day.

Here are a few cool-weather tips:

- Don't wet the mix any more than necessary.
- If a polyethylene vapor barrier isn't required, don't use one. Any moisture that drains out of the slab will speed the set.
- Pour as early as possible to avoid finishing the slab after dark.

Cold-Weather Pours

When the temperature is cold, a whole new set of rules comes into play. Concrete cannot be allowed to freeze. That tender, finely finished surface you just troweled on the slab will turn to mush if it's allowed to freeze. Fortunately, the chemical reaction that takes place when concrete hardens generates heat.

Here are some strategies that help in a cold-weather pour:

- Ask your concrete supplier about using warm mixing water to prevent problems during transit on days when the temperature is well below freezing.
- Having the supplier add calcium to the mix accelerates the initial set of the concrete, and the concrete achieves the strength to resist freeze/thaw stress faster. The amount of calcium is measured as a percentage of the cement content and ranges from ½ percent to 2 percent. Talk to a veteran concrete finisher before deciding when and how much calcium to add to the mix. Too much calcium produces the same problems as hot, dry weather. It's important to note that calcium is corrosive to steel and should never be used in steel-reinforced concrete.
- Always be sure that all components of the subbase are frost-free.
- Provide supplemental heat to keep the building above freezing.
- Cover the slab with polyethylene and then spread an insulating layer of straw or hay at least 4 in. thick on top, or use an insulating tarp.
- The best strategy: Pour when cold temperatures are not an issue.

Cement shoes. Kneeboards—pieces of plywood with strips of wood on one edge—allow you to move around on fresh concrete without sinking in because the boards distribute weight over a large surface area. Concrete finishing is done from kneeboards while the concrete is in a plastic state, meaning it's neither liquid nor solid.

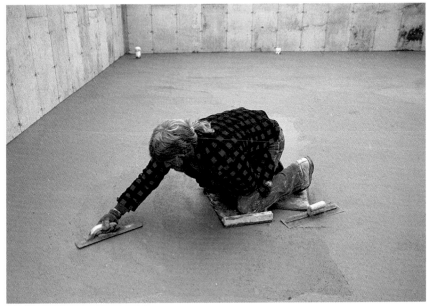

Start with a mag; finish with steel. After bull floating, use a magnesium float (top) to smooth out bumps and fill in low spots. The resulting finish will be coarse. Later, use a steel trowel (bottom) to get a smooth, dense finish that won't crumble when it's swept.

your first pass will be with a magnesium hand float (see the top photo above).

Like a bull float, a magnesium hand float works the fines to the surface, and you fill in any low spots or knock down any high spots during this pass. The goal when using the magnesium float is to level the concrete, preparing a surface that is ready for smoothing with the steel trowel. You can generally work the entire slab with the magnesium float before it's time to trowel with steel.

The difference between a magnesium float and a steel trowel is easy to recognize on the slab. You can work the slab all day long with magnesium, but you'll never get beyond a level, grainy surface. But when the slab is ready, and you lay a steel trowel to it, the results are impressive.

Hit the Slab with Steel

Keeping the image in your mind of the slab drying from the bottom up, picture the top ⅛ in. of the concrete, which is all cement, sand, and water. While this top section is in a plastic state—neither liquid nor solid—the steel trowel will smooth this layer and compact it into a dense, hard finish. Now the preparatory work pays off; if the concrete was placed and leveled accurately, the final finish goes quickly.

Obtaining an exceptionally smooth finish is a practiced technique that takes years to develop. The steeper the angle of the trowel to the slab, the more trowel marks will occur. If you hold the trowel at an extremely slight angle, you're liable to catch the slab and tear out the surface.

Your troweling technique will be dictated by how loose or tight the surface of the slab is. When the surface is wet, you can hold your trowel fairly flat, but as the fines tighten up, you'll have to increase both the angle and pressure of your trowel. As the slab dries you might have to use both hands on the trowel to muscle some fines to the surface (see the photos at right). Once the fines have emerged, switch back to one hand and polish the area with your trowel (see the bottom right photo on the facing page). If you've waited too long, and you're losing the slab, sprinkle water on its surface to buy a little more finishing time. After that, there isn't enough angle, pressure, or water anywhere on earth to bring a lost slab back to life. If it's important that the final finish be first rate, consider hiring a professional. Remember, you get only one try.

Curing the Finished Slab

While it's true that you can walk on the floor the day after it's placed, concrete actually hardens very slowly. The initial set represents about a quarter of the total strength; it takes about a month for concrete to cure fully. The goal during this period is to have the concrete cure as slowly as possible.

Keeping the slab soaked with water for four or five days will keep it from drying too quickly, but continual hosing down involves a lot of time and effort. Slabs can require a soaking every half hour in the heat of summer. A masonry sealer applied the day after the pour will keep the slab from drying too quickly and protect the floor from stains that might otherwise wick into the slab.

Two hands! Two hands! As the concrete sets up, working it with a steel trowel may require the strength of two hands. The back of the trowel is angled up as you push it away (top), and the front of the trowel is angled up as you pull it toward you (bottom).

When you consider that the material cost of a basement slab is less than $1 per sq. ft., it's difficult to imagine a more economical finished floor system. But when you consider the cost of removing and replacing an improperly finished concrete floor, the importance of knowing how to handle two or three truckloads of concrete becomes apparent.

Carl Hagstrom is a builder in Montrose, Pennsylvania, and vice-president at WOODWEB.com, an online resource for the professional woodworking industry.

Building a Block Foundation

■ BY DICK KREH

Foundation walls are often the most neglected part of a structure. But they are actually the most structurally important element of a house. They support the weight of the building by distributing its entire load over a large area. Apart from structural requirements, foundations have to be waterproofed, insulated, and properly drained.

Although the depth of a foundation wall may vary according to the specific needs of the site or building, the footings must always be below the frost line. If they're not, the foundation will heave in cold weather as the frozen earth swells, and then settle in warm weather when the ground softens. This shifting can crack foundations, rack framing, and make for wavy floors and sagging roofs.

Concrete blocks are composed of portland cement, a fine aggregate, and water. They have been a popular choice for foundations because they're not too expensive, they go up in a straightforward way, and they're available everywhere. Block foundations provide adequate compressive strength and resistance to fire and moisture. They don't require formwork, and they're not expensive to maintain.

All standard blocks are 8 in. high and 16 in. long—including the usual ⅜-in.-thick mortar head and bed joints. But they come in different widths. The size given for a block always refers to its width. The size you need depends on the vertical loads and lateral stresses that the wall will have to withstand, but as a rule, most concrete-block foundations are built of 10-in. or 12-in. block.

Footings

After the foundation area has been laid out and excavated, the concrete footings are poured. Footings should be about twice as wide as the block wall they will support. A 12-in. concrete-block foundation wall, for example, should have a 24-in.-wide footing. The average depth for the footing, unless there is a special problem, is 8 in.

Concrete footings for homes or small structures need a compressive strength of 2,500 pounds per square inch (psi). You can order a footing mix either by specifying a five-bag mix, which means that there are five bags of portland cement to each cubic yard, or by asking for a prescription mix—one that is ordered by giving a psi rating.

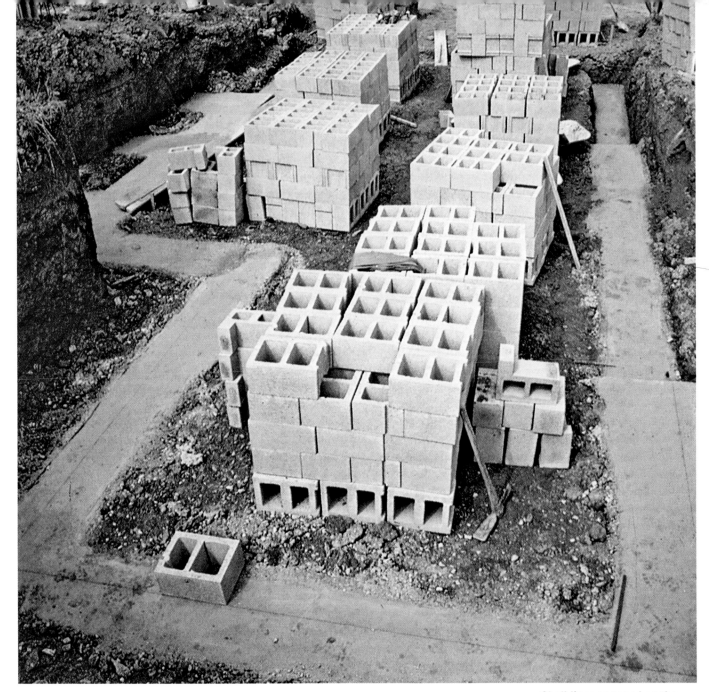

Chalklines snapped on the cured footing guide the masons in laying up the first course of concrete block for a foundation wall. The blocks are stacked around the site to minimize legwork, yet allow the masons enough room to work comfortably.

Some architects and local building codes require you to state the prescription mix when you order. Either way, footing concrete is a little less expensive than regular finishing concrete, which usually contains at least six bags of portland cement to the cubic yard. The six-bag mix is richer and easier to trowel, but isn't needed for most footings.

TYPES OF FOOTINGS

There are two types of footings—trench footings and formed footings. If the area where the walls are to be built is relatively free of rock, the simplest solution is to dig a trench and use it as a form. Keep the top of the concrete footings level by driving short lengths of rebar to the proper elevation. Don't use wooden stakes because later they'll rot and leave voids in your footing. You'll need a transit level or water level to get the rods at the right height. After you install the level rods even with the top of the proposed footing, pour concrete in the trench, and trowel it flush with the tops of the rods. Some building codes require that these stakes be removed before the concrete sets up.

If the ground is rocky, you may have to set up wooden forms and brace them for the pour. I've saved some money in this situation by ordering the floor joists for the first floor and using them to build the forms. This won't damage the joists and will save you a lot of money. When the concrete has set, I remove the boards and clean them off with a wire brush and water. The sooner you remove the forms, the easier it will be to clean them.

FIND THE CORNER POINTS

After the footings have cured for at least 24 hours, drive nails at the corners of the foundation. To find the corner points, use a transit level or drop a plumb line from the layout lines that are strung to your batter boards at the top of the foundation. Next, snap a chalkline between the corner nails on the footings to mark the wall lines. Stack the blocks around the inside of the foundation. Leave at least 2 ft. of working space between the footing and your stacks of block. Also, allow room for a traffic lane so the workers can get back and forth with mortar and scaffolding.

Mortar Mix for Block

For the average block foundation, use masonry cement, which is sold in 70-lb. bags. You have to supply sand and water. Masonry cement is made by many companies. Brand name doesn't matter much, but you will need to choose between mixes of different strength. The average strength, for general masonry work, is universally classified as Type N. Unless you ask for a special type, you'll always get Type N. I get Type N masonry cement unless there is a severe moisture condition or stress, in which case I would use Type M, which is much stronger. The correct proportions of sand and water are important to get full-strength mortar. Like concrete, mortar reaches test strength in 28 days, under normal weather conditions.

To mix the mortar, use one part masonry cement to three parts sand, with enough water to blend the ingredients into a workable mixture. Mortar for concrete block should be a little stiffer than for brickwork, because of the greater weight of the blocks. You will have to experiment a little to get it right. The mortar must be able to support the weight of the block without sinking.

The mixing water should be reasonably clean and free from mud, silt, or organic matter. Drinking water makes good mortar. Order washed building sand from your supplier. It's sold by the ton.

HOW MUCH MORTAR?

The following will help you estimate the amount of mortar you'll need: One bag of masonry cement when mixed with sand and water will lay about 28 concrete blocks. Eight bags of masonry cement, on the average, will require one ton of building sand. Remember that if you have the sand dumped on the ground, some will be lost since you can't pick it all up with the shovel. For each three tons, allow about a half-ton for waste.

Laying Out the First Course

Assuming the footing is level, begin by troweling down a bed of mortar and laying one block on the corner. Tap it down until it is the correct height (8 in.), level, and plumb.

A block wall built of either 10-in. or 12-in. block requires a special L-shaped corner block, which will bond half over the one beneath. The point is to avoid a continuous vertical mortar joint at the corner. Now lay the adjoining block. It will fit against the L-shaped corner block, forming the correct half-bond, as shown in the drawing and photo on the facing page.

When the second L-shaped corner block is laid over the one beneath in the opposite direction, the bond of the wall is established. On each succeeding course the L corner block will be reversed.

Once the first corner is laid out, measure the first course out to the opposite corner. It's best for the entire course to be laid in whole blocks. You can do this simply by using a steel tape, marking off increments of 48 in., which is three blocks including their mortar head joints. Or you can slide a 4-ft. level along where a level, continuous course of block runs through from one corner to the other. Steps in footings should be in increments of 8 in. so that courses of block work out evenly.

MAKE A STORY POLE

After one course of block is laid completely around the foundation to establish the bond and wall lines, it's time to build up the four corners. But before you begin laying block, you should make a story pole, sometimes called a course rod. Do this by selecting a fairly straight wooden pole and marking it off every 8 in. from the bottom to the height of the top of the foundation wall.

Any special elevations or features, such as window heads, door heads, sills, and beam pockets, should be marked on the pole to coincide with the 8-in. increments wherever possible. After checking all your pencil marks, make them permanent by kerfing the pole lightly with a saw. Then cut the pole off even with the top of the foundation wall and number the courses of block from the bottom to the top so you don't find yourself using it upside down.

BEGIN LAYING UP THE CORNERS

Now you can start laying up the corners so that you end up with only one block at the level of the top course. Successive courses are racked back half a block shorter than the previous ones (see the left photo on p. 80), so trowel on only enough mortar to bed the blocks in a given course. If the local code or your specifications call for using wire reinforcement in the joints, leave at least 6 in. of wire extending over the block. At the corners,

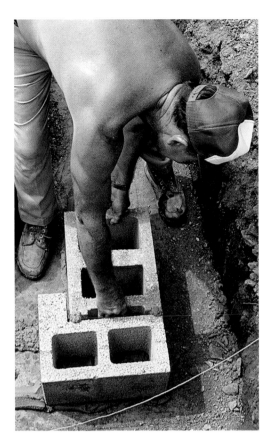

Laying the first course begins at the corners. Once the four corners are laid and aligned, the entire bottom course is laid directly atop the footing. Most foundation walls are built from block 10 in. or 12 in. wide, and special L-shaped corner blocks, like the one shown here, have to be ordered. Only 8-in.-wide blocks can be laid up without L-shaped corner blocks.

Building Up a Corner with L-Blocks

11⅝ in.

7⅝ in.

15⅝ in.

After the first course is laid, the corners are built up to the topmost course. Above, a mason checks course heights against a story pole, which is graduated in 8-in. increments.

Reinforced concrete lintels, right, are used to tie the main foundation to walls that are laid at a higher level, such as porch foundations or garage walls.

cut one strand of the wire, and bend the other at 90 degrees, rather than butting two sections together and having a break in the reinforcement.

Check the height of the blockwork periodically with the story pole. The courses of block should line up even with the kerfs. Once the corners are laid up, you can begin to fill in the wall between. Keep the courses level by laying them to a line stretched between the corners. Keep the corners plumb by checking every course with a spirit level.

Using Manufactured Corner Poles

So far, I've described laying a foundation using the traditional method of leveling and plumbing. But in recent years, manufactured metal corner-pole guides have become popular with builders. They guide the laying up of each course and require less skill than the old way. They work like this: The corner poles are set on the wall once the first course of block is laid out. They are plumbed, then braced in position. Each pole has course heights engraved on it. Line blocks are attached to the poles on opposite corners at the desired course height, and the wall is laid to the line. There is no doubt that the use of manufactured corner-pole guides has increased the mason's productivity without adversely affecting the quality of the work.

If you have to tie a porch or garage wall into a main foundation at a higher elevation, lay a concrete-block lintel in mortar from the corner of the wall built to the footing at the higher elevation (see the right photo on the facing page). Then lay blocks on the lintel to form the wall. This saves time and materials in an area that doesn't require a full-basement foundation.

Stepping the Wall at Grade Line

As you build up the natural grade line of the earth, you can set the front of the wall back about 4 in. to form a shelf for a brick veneer, if the plans call for it. This is done by switching to narrower block—from 12-in. block to 8-in. block or from 10-in. block to 6-in. block. The inside of the wall stays in the same plane.

Making the Last Course Solid

On some jobs, specifications require that the last course of block be solid to help distribute the weight of the structure above and to close off the holes. You need only grout the voids in the top course of block. Broken bits of block wedged into the voids in the course below will keep the concrete from falling through. The sill plasters will rest on this top course, and the floor joists on top of the plate.

The sill plate has to be bolted down to the top of the foundation wall. So you have to grout anchor bolts into the top of the wall every 4 ft. or 5 ft. These bolts should have an L-bend on the bottom and be

mortared in fully so they don't pull out when the nut is tightened against the sill plate. They should extend about 2 in. out of the top of the wall. In some parts of the country, building codes require that the walls include a steel-reinforced, poured-in-place concrete bond beam in every fourth course.

Waterproofing the Foundation

The traditional method of waterproofing a concrete block foundation is to parge (stucco) on two coats of mortar and then to apply a tar compound on top of that. This double protection works well unless there is a severe drainage problem, and the soil is liable to hold a lot of water for a long time.

MORTAR MIXES FOR PARGING

There are various mortar mixes you can use to parge the foundation. I recommend using a mix of one part portland cement to one-half part hydrated lime to three parts washed sand. This is a little richer than standard masonry cement and is known as type S mortar. The mix should be plastic or workable enough to trowel on the wall freely. Many mortars on the market that have waterproofers in them are all right to use. However, no two builders I know seem to agree on a mix, and most have worked out their own formulas.

PREPARE THE FOUNDATION WALL

Prepare the foundation wall for parging by scraping off mortar drips left on the block. Next, dampen the wall with a fine spray of water from a garden hose or a tank-type garden sprayer. Don't soak the wall, just moisten it. This prevents the parging mortar from drying out too quickly and allows it to cure slowly and create a better bond with the wall.

Start parging on the first coat from the bottom of the wall to the top, about ¼ in. thick. A plastering or cement-finishing

> *The use of manufactured corner-pole guides has increased the mason's productivity without adversely affecting the quality of the work.*

Troweling Technique

Laying up concrete blocks with speed and precision takes a lot of practice. But it's chiefly a matter of learning several tricks, developing trowel skills, and performing repetitive motions for several days. A journeyman mason can lay an average of 200 10-in. or 12-in. blocks in eight hours. A nonprofessional, working carefully and after practicing the techniques shown here, ought to be able to lay half that many. If you've never laid block before, what follows will show you the basic steps involved in laying up a block wall.

1. First, mix the mortar to the correct stiffness to support the weight of the block. Then apply mortar for the bed joints by picking up a trowelful from the mortarboard and setting it on the trowel with a downward jar of the wrist. Then swipe the mortar onto the outside edges of the top of the block with a quick downward motion, as shown.

2. Apply the mortar head joint pretty much the same way. Set the block on its end, pick up some mortar on the trowel, set it on the trowel with a downward jerk, and then swipe it on the top edges of the block (both sides).

3. After buttering both edges of the block with mortar, press the inside edges of the block with mortar, and press the inside edge of the mortar in the head joints down at an angle. This prevents the mortar from falling off when the block is picked up and laid in the wall.

4. Lay the block on the mortar bed close to the line, tapping down with the blade of the trowel until the block is level with the top edge of the line. Tap the block in the center so you won't chip and smear the face with mortar. Use a hammer if the block does not settle easily into place.

5. The mortar in the head joint should squeeze out to form a full joint at the edge if you've buttered it right. The face of the block should be laid about $\frac{1}{16}$ in. back from the line to keep the wall from bowing out. You can judge this by eye-balling a little light between the line and the block.

6. Remove the excess mortar that's oozing out of the joint with the trowel held slightly at an angle so that you don't smear the face of the block with mud. Return the excess mortar to the mortarboard.

Check the height of the blockwork by holding the story pole on the base and reading the figure to the top of the block. Courses should be increments of 8 in.

Finishing the Joints

Different types of joint finishes can be achieved with different tools. The most popular by far is the concave or half-round joint, which you make by running the jointing tool through the head joints first, and then through the bed joints to form a straight, continuous horizontal joint. If you buy this jointing tool, be sure that you get a convex jointer. These are available in sled-runner type or in a smaller pocket size. I like the sled runner because it makes a straighter joint.

After the mortar has dried enough so it won't smear (about a half-hour), brush the joints lightly to remove any remaining particles of mortar.

The completed foundation has been sealed with two ¼-in.-thick parging coats and topped with an application of tar compound, which finishes the waterproofing. Backfilling should happen only after the first floor is framed and the walls framed up, so the added weight of the structure will stiffen the walls and make them less liable to bulge from the pressure of the earth.

trowel is excellent for this. After troweling on the parging, scratch the surface with an old broom or a tool made for this purpose. Let the mortar dry for about 24 hours or until the next day, and repeat the process for the second coat. Dampen the wall between coats for a good bond. Trowel the final coat smooth, and let it dry for another 24 hours.

To complete the waterproofing job, spread on two coats of tar compound (see the photo above). You can do this with a brush or roller if the weather is warm. Many builders in my area use a product called Hydrocide® 700B from Sonneborn Building Products. It comes in 5-gal. containers and is available from most building-supply dealers. I like it because it stays a little tacky and seals the wall very well. It's gooey, though, so wear old clothes and gloves when you're applying it. Kerosene will get it off your hands and tools when the job is done.

Drain Tile

Most codes require some type of drain tile or pipe around the foundation to divert water buildup and to help keep the basement dry. The design of the drain-tile system is important. Generally, drain tile or pipe is installed around the exterior wall of the foundation, below the wall but above the bottom of the footing.

Begin by spreading a bed of crushed stone or gravel around the foundation next to the wall. Lay the drain tile or perforated plastic pipe on top of this bed. The bottom of the drain pipe should never be lower than the bottom of the footing, or it won't work properly. Lay filter cloth over the drain pipe to keep mud and dirt from blocking the holes. Then place another 4-in. to 6-in. layer of crushed stone or gravel over the pipe, as shown in the drawing on the facing page.

The water collected by the drain tile has to flow away from the foundation. One way to make this happen is to drain the water to a natural drain away from the foundation

Section of Foundation Wall and Footing

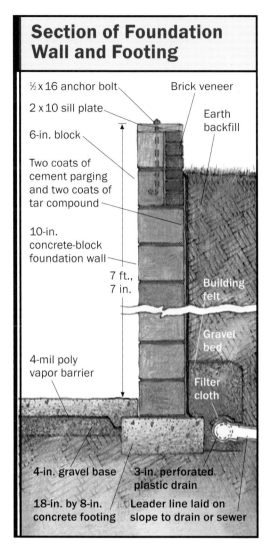

- ½ x 16 anchor bolt
- Brick veneer
- 2 x 10 sill plate
- Earth backfill
- 6-in. block
- Two coats of cement parging and two coats of tar compound
- 10-in. concrete-block foundation wall
- 7 ft., 7 in.
- Building felt
- Gravel bed
- Filter cloth
- 4-mil poly vapor barrier
- 4-in. gravel base
- 3-in. perforated plastic drain
- 18-in. by 8-in. concrete footing
- Leader line laid on slope to drain or sewer

area by installing a leader line on a slope that is lower than the drain pipe. The other method is to drain the water under the wall of the foundation and into a sump pit inside the basement. The water that collects is then pumped into a pipe up to grade or street level and allowed to drain away there naturally.

A third method, which has worked well for contractors in my area, is to put the drain tile inside the foundation on a bed of crushed stone, just beneath the finished concrete floor, which will be poured after the drain tile is in place. One-inch plastic pipe is installed about 6 ft. o.c. through the wall at the bottom of the head joints in the first course of block. When the foundation wall is done, crushed stone is spread around

the exterior edges as before, but no drain pipes are needed.

The idea is that any water that builds up outside the foundation wall will drain through and into the drain tiles. In addition, the water inside the foundation area will also flow into the drain tile and into the sump pit in the basement door, where it can be pumped out to grade level to drain away.

Insulating the Foundation

The use of rigid insulation applied to the exterior of the foundation wall has helped to reduce dampness and heat loss. This is especially important in the construction of earth-sheltered homes. There are a number of products that will do a good job. Generally, rigid insulation is applied to the waterproofed wall with a mastic adhesive that's spread on the back of the foam board. Use a mastic or caulking that does not have an asphalt base. Most panel adhesives will work. The building-supply dealer who sells the insulation will know the proper adhesive to use for a specific type of insulation board. Also, there are granular and other types of insulation that can be poured into the block cells.

After all the foundation work has been completed, the backfilling of earth should be done with great care so that the walls don't get pushed out of plumb. It is always better to wait until the first floor is framed up before backfilling. This weight resting on the foundation helps prevent cracking of the walls, and the framing material will brace the block wall and make it more rigid. If the walls are cracked or pushed in from backfilling, the only cure is very expensive—excavate again and replace the walls.

Dick Kreh is an author, mason, and industrial-arts teacher in Frederick, Maryland, and the author of Building with Masonry *(The Taunton Press, 1998).*

After the foundation work has been completed, the backfilling of earth should be done with great care so that the walls don't get pushed out of plumb.

Laying Up Concrete Block

■ BY JOHN CARROLL

Masons work from the ground up. You're probably not surprised to hear that. As simple as this principle sounds, however, it's also a bit misleading. Masons build from the ground up, but they measure from the top down, which means the real starting point of a masonry job is the top line.

From this line, masons measure down, marking in equal increments what will be the top of each course of brick or block. By working to these marks, masons arrive with evenly spaced courses at the tops of foundations, the bottoms of windowsills, or other predetermined landing places.

Even when I'm using basic techniques for planning and laying up a simple concrete-block foundation, I return again and again to the same starting point: the top line.

Planning and Laying Out a Block Structure

Before any mud is slung or any blocks are buttered—indeed, before any dirt is dug or any concrete footings are poured—the finished dimensions of a block structure should be established. These dimensions are length and width—along with any variations in the basic rectangle—and exact height.

When I build a foundation, for instance, the first thing I do is see if any small dimension adjustments might allow the masonry units to fit without being cut. Say this foundation is 12 ft. by 13 ft. I try to change the 13-ft. dimension to 13 ft. 4 in. so that it works out to an even number of blocks.

Unlike the foundation for a freestanding building, where a difference of an inch or two in the final height usually isn't too important, the height of the foundation for an addition must be right on the money. To get to this point, the first thing I do is to mark the height of the existing finished floor on the outside wall of the house. For this, I set up my laser level, shoot the elevation of the floor through an open door or window, then transfer that elevation to the outside wall.

Next, I find out exactly what will be used in the addition as floor covering, subfloor, joists, and sills. On the foundation project shown here, these included carpet and pad (¾ in.), tongue-and-groove plywood (¾ in.), 2x10 joists (9¼ in.), and 2x8 sills (1½ in.). By measuring down a total of these depths—12¼ in.—from the mark I made on the out-

Measure from the top down. Steel story poles at the outside corners of the block foundation are nailed to the footings and held in place with 1x4 braces. They mark the corners of the foundation and the courses of block. Strings between the house and the poles mark the foundation height.

The basic tools for working with concrete block. To get started laying up block requires a trowel, a pointing tool, line blocks, and string. Also helpful are line stretchers, left, and twigs, shown next to the string.

side wall, I arrived at the correct height for the top of the foundation. Using my laser level, I made marks at this elevation at both ends of the planned addition, and I snapped a line on the house to represent the top of the new foundation.

Establishing the Entire Top-of-Foundation Line

The project pictured here is a 16-ft. by 24-ft. foundation for an addition. Here's how it began. After I snapped a line on the house to represent the top of the foundation, I marked the beginning and end of the foundation on this chalkline, which served as a reference to lay out the rest of the foundation.

My next step was to set up a line that would represent the top of the outside foundation wall. This line had to be 16 ft. from—and parallel to—the house and at the same level as the chalkline. To hold the string, I set up a pair of batter boards. First, I drove a pair of 2x4 stakes into the ground several feet outside the corners of the planned foundation (I could judge this by eye by looking at the corner marks on the chalkline marked on the house).

I set the stakes in line, 14 ft. and 18 ft. out from the house so that they would straddle the planned 16-ft. line. After driving in the 2x4s, I used my laser level to transfer the top-of-foundation elevation, represented by the chalkline, to the stakes. Then I attached a horizontal batter board to each pair of stakes, keeping the top edge even with the top-of-foundation marks. Once the batter boards were attached, I stretched a line from one batter board to the other. To hold the line in place, I used mason's line blocks, which I could slide along the batter board. By methodically measuring and adjusting the work I was doing, I was able to get the line exactly parallel to and 16 ft. away from the house. This string represented the top outside edge of the foundation wall.

I now turned to the task of laying out the two sidewalls. These walls would run perpendicular to the house and would begin at the corners marked on the chalkline. To do this, I calculated the hypotenuse of a right triangle with sides of 16 ft. and 24 ft. This works out to 28 ft. 10⅛ in. I pulled this dimension diagonally across the planned foundation from one of the corner marks on the house to the string. Using a felt-tip pen, I marked where 28 ft. 10⅛ in. intersected the string. This dot represented the third corner of the foundation. Pulling this dimension diagonally in the other direction, I marked the fourth corner. To check my work, I measured from dot to dot along the string. Seeing that it was exactly 24 ft., I knew my layout was correct.

After the third and fourth corners were marked on the string, I set up two more pairs of batter boards. To represent the top of each wall, I stretched a string from the corner mark on the house, through the mark on the string, to the new batter board. I now had strings outlining the top outside edge of the entire foundation.

A grinder makes a clean cut. To avoid ragged edges in cut blocks, use a grinder equipped with a diamond blade and give the block a tap with a hammer to get a clean split. This procedure is messy, however, so a respirator and eye protection are important.

Tap the block gently to get it plumb and level. This block was set into about an inch of mortar and gently tapped with the butt of the trowel until it was level and plumb. The layout line remains a fraction of an inch above the block to keep the string from snagging on the block.

Not too mushy and not too stiff. Mortar should be as wet as possible, yet still be workable. Mortar should have some body, but it still should be mushy.

You can feel if a block is aligned with the rest. Place the palm of your hand on the new block and your fingers across the joint between blocks to feel if the new block is even.

A Story Pole Aligns the Courses and the Corners

The top of the footing should be a set number of block courses below the top-of-foundation line. In the example that I'm using, the footing was exactly 80 in., or 10 block courses, below the top line.

After the footings are poured, it's good to recheck the position of the top-of-foundation lines. After seeing that my top-of-foundation line hadn't moved, I dropped a plumb bob from the intersection of the lines to mark the outside corners of the block courses. At each outside corner, I attached a story pole of 2-in. tubular steel to the footing. The story poles I use have two L-shaped brackets welded to their bottoms. To attach the story poles, I nail through holes in the brackets with case-hardened, fluted masonry nails into the concrete footing. Then I set the story pole plumb, and I clamped 1x4 braces to the pole and to the batter boards.

At the corners that engage the house, I affixed 1x4 story poles. Next, I marked down all the story poles in 8-in. increments from the top-of-foundation line and was ready to lay block.

For Efficiency and Looks, Avoid Cut Blocks

Blocks are difficult and expensive to cut, and when they are cut they can detract from the appearance of the job. So it's wise to avoid unnecessary cuts. Cutting units lengthwise (the masonry equivalent of ripping) is referred to as splitting. Cutting units to length (the masonry equivalent of cross-cutting) is simply called cutting. Split blocks not only interrupt the orderly progression of horizontal courses—and look bad—but they're also a scourge of productivity. Not surprisingly, masons do their utmost to avoid split units. Units cut to length are a different matter. Because every house has windows, doors, and corners, blocks inevitably have to be cut to length.

When I have to cut block, I avoid masonry hammers or chisels. I've never liked the ragged edges these tools leave. Instead, I use a 4-in. grinder equipped with a diamond blade (see the left photo on p. 89). You can also use an abrasive blade in a circular saw. After cutting as far into the block as I can, I give it a good tap with a hammer and it breaks easily.

Buckets Are Better Than Shovels When Mixing Mortar

Block work is mainly a physical skill that takes a lot of practice, ideally alongside an experienced mason.

The first thing a novice should learn is how to make good, consistent mortar. I've found that this cannot be done using ready-mix mortar. The material that comes already mixed with sand has poor plasticity and contains coarse sand.

For mixing mortar, you can rent electric mortar mixers for between $30 and $50 a day. However, most novices are laying block so slowly that they'd be just as well off mixing it in a wheelbarrow by hand.

Mortar is made up of portland cement mixed either with lime or any of several proprietary ingredients. One part of this mixture is combined with 2½ parts to 3 parts sand, and water is added to get the right consistency. Cement/lime mortars often come in separate bags and have to be mixed on the job. Masonry-cement mortars (those with the proprietary ingredients) come ready to be mixed with sand.

Masons often count 18 shovels of sand for every bag of mortar, but the size of a shovel is inexact. I measure the ingredients by filling a drywall bucket with dry mortar and three other drywall buckets with sand. I put all this in a mixer and add water until I get the right consistency.

It's hard to describe the right consistency. Unlike concrete, which should be kept as stiff as possible, mortar should be made as wet as possible, yet still be workable. The primary role of mortar is to bond masonry units together. Wet mortar spread on dry units is absorbed deep into the pores and crevices of the units, producing a tenacious bond. A mixture that is too wet, however, is almost impossible to work with and makes a mess of the job. Good mortar is almost fluffy; some masons call mortar that's just right "fat mortar."

The best way to learn how to make good, wet, workable mortar is by actually making it and using it. Even perfect mortar doesn't stay that way for long. On hot days, you often have to "shake up" the mortar by mixing in a little water. The longer mortar sits, the harder it is to work. After two hours, it should be disposed of, usually into the cells of the block wall.

I try to make about an hour's worth of mortar at a time. Usually one bucket of mortar and three buckets of sand last me about an hour; however, a novice may want to start by mixing a half-batch. And before I mix the material, I make sure I have everything ready. To estimate what you'll need, figure three bags of mortar for every 100 blocks plus 9 cu. ft. of sand, or about fourteen 5-gal. buckets (a 5-gal. bucket is 0.668 cu. ft.).

Consistent Mortar Joints Testify to Good Work

Masons adjust the height or length of courses by altering the thickness of mortar joints. This is a basic part of masonry. It allows masons to make up for inconsistencies in the size of the units and to fit whole units into a given space (between windows, for instance). But this device is easily overused.

TIP

At the end of the day, it's smart to cover the top of the wall, both to protect the bond of the day's work and to make sure the top course is dry when you return to lay blocks on top of it.

Clean the joint of excess mortar. After a block is set in place, use the trowel to cut the mortar that has bulged out. Then give the trowel a shake and butter the end of the block with the remaining mortar.

Fat joints and abrupt changes in the thickness of joints look terrible.

Joints that are thicker than ¾ in. can shrink excessively, which sometime results in leaky hairline cracks. And because masons tend to use a stiff mix when making thick joints, the bond is often poor, resulting in both leakage and structural compromise. For optimal appearance and performance, joints should be between 5/16 in. and ½ in. thick, and they should be as consistent as possible.

To attain evenly spaced courses with consistent mortar joints, masons lay out courses in standard modules based on the size of the block used. The height increments for concrete blocks are typically 8 in. or 4 in. Like lumber, blocks are smaller than their nominal size; an 8-in. block is actually 7⅝ in., and a 4-in. block is 3⅝ in. A block

with a ⅜-in. bed joint, then, measures an even 8 in. or 4 in.

Getting Started with the First Course

After stretching stringlines for the first course in both directions from the corner pole, I lay my first block, a corner block. The top of this block fits snugly to the intersection formed by the strings.

After getting this block even with both strings, I start laying blocks in from the corner along one of the lines. After laying all or part of this line, I work down the other line, again working from the corner in.

After running all or part of this line, I move both lines up to the next mark on the story pole and start the second course. Like the first block I laid in the previous course, the first one here fits snugly into the intersection of the lines at the story pole. This block, however, crosses to the corner block in the first course and thus begins the familiar bond pattern associated with most block walls.

At this point in the job I have the option of either laying the entire length of each course or laying just far enough down the line to build up the corner. Usually, there's no need to build a corner lead. I often build entire small foundations straight off the story poles. On congested sites, however, it's often best to build the corner first so that you can get the story pole and braces out of the way as soon as possible. To set up a line for laying the rest of the wall, just hook the line block directly to the corner you've built.

The ability to set up a line quickly and securely is an important masonry skill. The two most common tools for doing this are line blocks and twigs, (see the bottom photo on p. 88) which masonry-supply houses traditionally supply for free. To use a line block, wrap the string over and around the block a couple of times and hook it on either the outside corner of the masonry or on a corner pole.

A twig fastens to the string, and a brick or a piece of block holds it on the top edge of the block. One of the advantages of using a twig is that the string can be pulled taut without exerting pressure on the masonry just laid. Sometimes, instead of hooking a line block on a corner I've laid, I attach the line to a stake or a block beyond the corner and use a twig to hold the line even with the top edge of the unit.

A third tool that is sometimes used to affix lines to the inside corners is the line stretcher. After the corner has been built, the string is wrapped around the line stretcher, and the stretcher is placed across the top of the first unit of the course being laid. The stretcher is held in place by tension on the line. All these tools are available from a good masonry-supply company.

While you're pulling together the tools and materials you need to get started, you should also consider lightweight block instead of standard block. Lightweight block is made with special lightweight aggregates rather than crushed stone. A standard 8-in. by 8-in. by 16-in. block weighs 32 lb. to 38 lb. A lightweight block weighs only 22 lb. to 27 lb., but it's weight that really adds up after a day of lifting. Lightweight block also costs up to 25 percent more than normal-weight block.

Troweling Techniques: To Sling or to Butter

There are two techniques for spreading the bed joints, or horizontal joints, of a block wall. In the first, the mason loads his trowel, holds it above the top edge of the lower block and shakes some mortar loose from the trowel. Then, as he moves the trowel down the line toward himself, he shakes more of the mortar loose and lets it fall in a line along the block. After about a third of the mortar is laid down like this, he simultaneously turns his wrist downward and pulls the trowel quickly toward himself, slinging the rest of the mortar in a nice line on top of the block.

In the second technique, the mason loads his trowel and gives it a good shake (a hard shake makes the mortar stick to the trowel). Then, in a downward, pulling stroke, he butters the top edge of the block by sliding the trowel blade down and across the block.

There are also two methods for buttering the head joints of blocks. Some masons butter the ends of the block before they pick it up and set it in the wall. Others butter the ends of the block after they set it in the wall. Either way, you need to apply the mortar with some oomph so that it hits the surface hard and sticks.

Finally, a few words on the trowel, which is the most basic tool in masonry. For me, the size of a trowel is analogous to the size of a hammer. Some carpenters use a 24-oz. hammer, and some masons use a 13-in. trowel.

Good mortar is almost fluffy; some masons call mortar that's just right "fat mortar."

"Ladder" reinforcement equals lateral strength. Ladder-type wire reinforcement can be added to the bed joints of a block wall as it's laid up. In this example, the ladder reinforcements were added at every other course.

A small trowel is 10½ in., and a large trowel is 13 in. My suggestion is to use whatever is most comfortable in your hand. I use a small hammer and a small trowel because I have tendonitis. So whatever you use depends on how much weight you want to handle.

After a While, You'll Get into the Rhythm of Working with Mortar and Block

Once I get going, the basic rhythm of masonry is to spread a bed of mortar an inch or so high, then gently push or tap the block until it's even with the line.

I like to keep about a ¹⁄₁₆-in. space between the line and the block, because if the block touches the line it can push it out (see the top photo on p. 90) and throw off a course. I almost always set the line up on the far side of the wall so that I don't have to lift blocks over it all day. As I set each block in the wall, I use the heel of my left hand (I'm right-handed) to feel when the top of the block is even with the preceding block (see the bottom right photo on p. 90). I get this corner even by feel and, looking straight down over the line, push or tap the outside of the block even with both the string and the block below it. At the same time, I push the block horizontally against the buttered ends of the preceding block until the joint size looks right.

Tooling the joints for a neat look. When the mortar is as stiff as putty and beginning to pull away slightly from the edges of the blocks, it's ready for tooling.

Using the trowel in my right hand, I cut the mortar that has bulged out from the joints (see the left photo on p. 92) and, giving my trowel a good shake, butter the end of the block I just laid (see the right photo on p. 92). I do any minor adjustments immediately. After blocks have set for a while, they should not be disturbed. Tapping blocks laterally after the initial set of the mortar breaks the bond and weakens the wall.

Another important consideration in block work is keeping the blocks dry. Although dry and absorbent brick sometimes needs to be dampened before it is laid, blocks should always be kept dry before they are installed. Wet blocks don't bond well and are difficult to lay, so it's prudent to cover blocks as soon as they're delivered. At the end of the day, it's also smart to cover the top of the wall, both to protect the bond of the day's work and to make sure the top course is dry when you return to lay blocks on top of it.

To finish off the wall of the foundation, I install anchor bolts at every corner and at each door and window jamb, as the local code requires. I also install bolts at least every 6 ft. between these points. To anchor the sills, I drop broken bits of block or brick into the cores to provide support for the grout, which I make out of concrete or a portland cement–sand mix, which is stronger than mortar.

As for tooling the joints, it's necessary to wait until the mortar is thumbprint hard before tooling begins. When the joints are ready, they are usually about as stiff as putty and starting to pull away slightly from the edges of the blocks. Using a jointing tool, I tool the vertical joints first, then do the horizontal joints (see the bottom photo on p. 88). Wherever the joints need a bit of mortar, I push a lump into the joint with my jointer.

There are many different jointers on the market. I own several different types. But for laying block, the S-shape jointer is a standard. You don't get into much variation in jointers until you get into brickwork. The S-shape jointer has two different-size jointers, with one at each end. Use whichever end looks best. I usually use the widest one so that it doesn't cut the joint too deeply.

Sometimes a Block Wall Requires Reinforcement

There are several ways to reinforce a unit masonry wall. The wall can be thickened by switching from 8-in. to 10-in. or 12-in. blocks. Ladder-type wire reinforcement can be added in the bed joints as the blocks are laid up (see the photo on the facing page). You can apply stucco reinforced with fibers to the outside of the wall after it's built. Or you can reinforce the wall with rebar and concrete grout.

To design a reinforced-block wall, it's best to hire a structural engineer for the project. You can also get an idea of what works in the area where you live by talking to other builders and to your local building inspector. Also, you can see what doesn't work by carefully looking in your area at foundations and retaining walls that have failed.

John Carroll is the author of numerous Taunton Press books, including Measuring, Marking & Layout *(1999) and* Working Alone *(1999).*

Wet blocks don't bond well and are difficult to lay, so it's prudent to cover blocks as soon as they're delivered.

TIP

To estimate what you'll need, figure three bags of mortar for every 100 blocks plus 9 cu. ft. of sand, or about 14 5-gal. buckets (a 5-gal. bucket is 0.668 cu. ft.).

Forming and Pouring Footings

■ BY RICK ARNOLD AND MIKE GUERTIN

Over the years, we've built homes on almost every type of foundation imaginable. However, a concrete foundation always seemed to provide the best base for a home built in Rhode Island, our part of the country. In 1996, Rick bought a concrete-forms company, and our firsthand knowledge of footings and foundations increased exponentially.

With every house, we do everything in our power to keep the house and its foundation from settling and cracking, which can cause problems ranging from drywall cracks and sloping floors to doors that won't close. The best preventive medicine is putting the foundation on top of poured-concrete footings (see the sidebar on p. 101).

The price for this medicine is usually reasonable. On a 26-ft. by 38-ft. house, footings add only about $600 to $700 to the cost of the house. Prospective homeowners will spend that much in a blink to upgrade a kitchen. We figure that it makes more sense to upgrade the whole house by adding footings to ensure that the new kitchen stays put.

Getting the Footings in the Right Spot

Most of the footings called for in our work are a foot high and 2 ft. wide. We normally reinforce footings with a double row of ½-in. (#4) steel rebar unless plans specify otherwise. For the project featured in this article, the soil at the bottom of the excavated hole was like beach sand, so footings were a must. We went with standard-size footings to support the 10-in.-wide by 8-ft.-high foundation walls that would be poured on top.

But before we can think about footings, we have to lay out the location of the foundation. We begin by establishing two starting points, the corners at both ends of one foundation wall. With most houses, there are a couple of surveyor stakes (see photo 1 on p. 98) outside the hole with the offsets (the distance to the edge of the foundation wall) written on them.

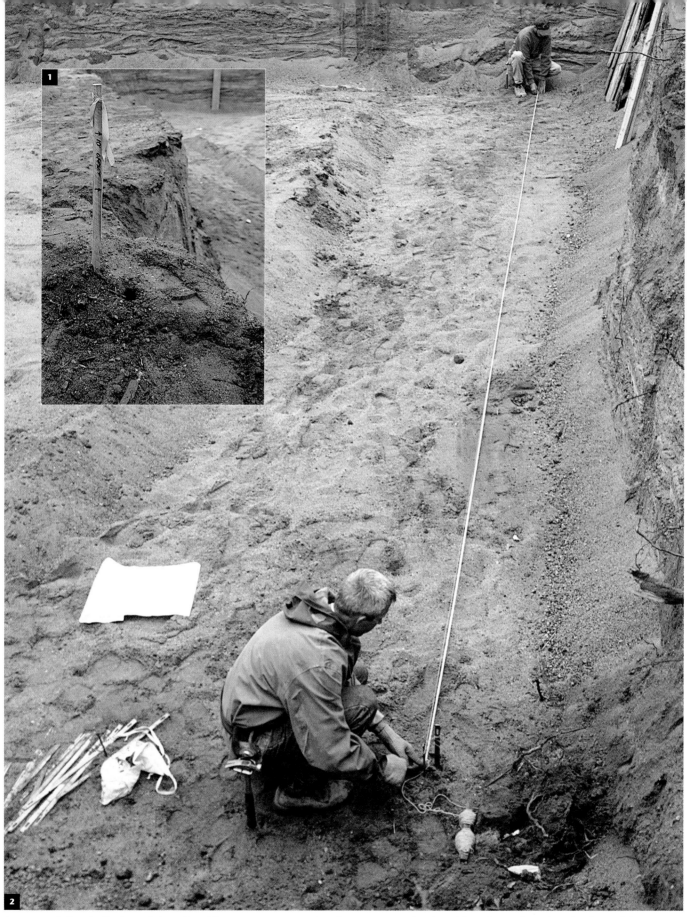

Surveyor stakes indicate wall location. Measurements are taken from surveyors' stakes (1) and transferred to the floor of the excavated hole to locate one foundation wall. The ends of the wall are then located precisely with a measuring tape (2).

For most jobs we find corner points by running a string between the stakes. We measure in the offset amount and then drop a couple of plumb lines to the floor of the foundation hole.

We drive stakes (usually foundation-form rods) into the ground at the two points; the measurement between the two rods should be the length of the wall as indicated on the plans (see photo 2 on the facing page). If the points are off an inch or two, we adjust them until the measurement is correct. If there is a large gap between our measurement and the plan, we call the surveyor back. For the house in this article, setbacks were tight, so the surveyor set two exact foundation corner points on the floor of the hole.

High-School Geometry and a Construction Calculator

The best way to locate the rest of the corners or points from the two reference points is to use diagonal measurements, just like we learned in high-school math. Unlike in high school, however, we depend on a calculator to do the math.

Before jumping into the hole, we sit down with the blueprints and the calculator. In about ten minutes we figure out our diagonals so that every corner on the plans has two reference measurements. Then it's just a matter of measuring those distances from the original two points to find the other foundation corners. The quickest method is to have two crew members hold tapes at the original reference points. A third crew member pulls the tapes tight and crosses them, moving the tapes until the calculated measurements from the corners meet each other (see photo 3 above). A stake is then driven into the ground to mark the point. We always double-check our measurements between the corners to make sure the foundation dimensions are right on.

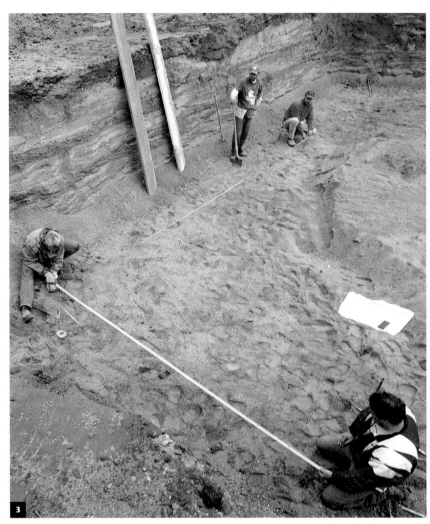

Diagonal measurements locate the other walls. All other corners are located by triangular measurements from the initial reference points (3). Three people and two tapes make the process go quicker.

The house in this article has a garage with a connecting breezeway that meets the house at an angle. To locate the corners of the breezeway, we triangulate from the foundation corners we've just found for the main body of the house. Again, we double-check all our dimensions to be sure they gibe with the plan. Once stakes have been driven at every corner, we run a string from one to the other, sort of like a giant connect-the-dots game. The result is an outline of the entire foundation.

> *We always double-check our measurements between the corners to make sure the foundation dimensions are right on.*

Outside forms are set up first. Starting at one corner, the outside forms are set at a given distance off the foundation line (4).

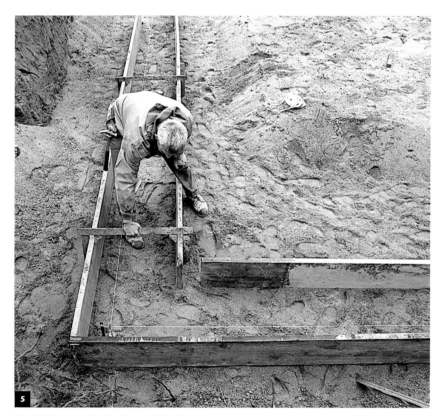

Brackets set the footing width. Brackets are then slipped over the forms that keep the inside form exactly 2 ft. away from the outside (5). The footing height is then set with a transit and marked on the inside of the form.

Footings Are Built for Strength and Function, Not Aesthetics

With the outline in place, we're ready to start setting forms. We use 2x12s connected by steel form brackets that hold the inside and outside forms exactly 2 ft. apart (see photo 5 at left). Rick had these brackets custom made, but there has been enough interest in them that he hopes to be offering them for sale soon.

We locate the forms so that the foundation walls will be centered on the footings. The foundation walls for this house were 10 in. wide, so we subtracted 10 in. from the footing width of 24 in. To leave equal amounts of footing on both sides of the foundation wall, we divided the remaining 14 in. in half to give us 7 in. of exposed footing on both sides of the walls.

Because the string we've set indicates the outside of the foundation wall, we start by nailing two 2x12s together to form an outside corner. We then set our corner 7 in. away from the string (see photo 4 on the facing page). A form bracket locks the inside form plank at the proper width (see photo 5 on the facing page). We continue this process around the perimeter of the foundation. Form brackets are dropped every 4 ft. or so. Where two planks butt, we toenail the tops together and put form brackets on both sides of the butt joint.

We keep many different lengths of form stock on hand, so we rarely need to cut a piece. When we come to a small jog, such as the inside corner of the angled breezeway, we simply form the whole area off. The extra concrete used is negligible. When two planks don't quite meet, we bridge the gap with a short piece, and if a plank is a little long, we just run it by the adjacent plank and tack the two together. Forming footings is forgiving because the finished product is buried. The major concern with footings is strength and function, not how they look.

When we need a footing that is wider than 2 ft. or if we're forming a large area such as a bulkhead pad, we secure the forms in a different manner. We run ¾-in.-wide perforated-steel strapping beneath the planks across the bottom of the footing. The strapping is run up the outside of the form planks and nailed to keep the bottoms of the planks from spreading. The tops are then held at the proper width with a length of 1x3 nailed between the two planks.

Footing Height Is Set with a Transit

Once all the forms are in place, we backfill against potential weak points such as butt joints or the larger areas where we couldn't use form brackets. Backfilling prevents the concrete from getting underneath the planks and lifting them up during the pour. At this point we transfer the foundation lines to the tops of the forms for future reference. The string is removed, but the stakes are left in the ground as a visual reference to make sure that the forms don't shift during the pour.

Next, using a builder's transit, we set the grade to make sure the top of the footing is poured to the same level. First, we find the lowest point on the forms by checking the height of the planks at every corner and at several points in between. The low point becomes the grade for the footing. We mark a yardstick to the measurement at the low point on the forms. Then we work our way back around the footing, installing grade nails every few feet on the inside of the forms. To mark the grade, we hold a 6d nail against the bottom of the yardstick and move it up or down until it's at the right height. The nail is then hammered in about halfway.

The excavators we work with usually leave the foundation hole less than 2 in. out of level. But if we see that some parts of the footing are too shallow because the bottom

The major concern with footings is strength and function, not how they look.

Why Footings?

Trudging through 2 ft. of snow can be a real nuisance. Strap on a pair of skis, though, and nuisance becomes recreation. The extra surface area of the skis distributes your weight over a greater area so that you don't sink into the snow. The same principle of load distribution is why we put footings under house foundations.

Footings are reinforced-concrete platforms at least twice as wide as the foundation they support and usually a foot deep. They are required by most local codes, especially if an area has soils with questionable bearing capacity, such as loose sand, silt, or clay. A wet site is another prime candidate for footings.

of the hole is too high after we shoot the grade, we dig out inside the forms until the proper depth is reached. If we undermine the planks in the process, we backfill the outside of the form. We also spray a light coat of form-release oil on the inside of the 2xs.

Let the Concrete Fly

Now we're ready for the concrete. The best foundation holes have clear access for the cement trucks all the way around the hole. Good access makes the pour go more quickly because the truck can shoot concrete into the forms without our having to drag it with shovels. If access is a problem, we usually start the pour at the most difficult spot for the truck to reach and work our way out.

The concrete chute is moved slowly along the forms, allowing the concrete to fill up to the grade nails. For sections that can't be reached with the truck's chute, we drag the concrete along the forms until those areas are filled. When the pour is finished, we begin installing a double row of ½-in. steel rebar around the footing (see photo 6 below).

Although rebar is not required for residential footings here in the Northeast, we believe that its added strength is cheap insurance. For commercial footings that are required to take many times the load of a typical house, we wire the rebar together and set it on chairs that keep it at a specific height during the pour. However, the additional cost of the labor and the materials usually rules out this option when we're doing residential projects.

Instead, rebar is placed atop the wet concrete about 6 in. in from each form. For angles or 90-degree corners, we bend the rebar around a knee until it's at the desired angle. Rebar is then inserted under the brackets and pushed down about 8 in. into the concrete, using the shovel as a gauge. As rebar is pushed in, we jiggle it with the shovel to remove any trapped air.

A 2x4 Cuts the Keyway

We level the concrete by vigorously pushing it with the flat of the shovel until the concrete is at finish grade. We add or remove a shovelful of concrete to adjust the level and to rework the concrete until the grade nails are exactly half-exposed.

After the concrete is leveled to the grade nails, we gently lift up all the form brackets a couple of inches, which makes our last two jobs easier. The first job is troweling the top of the footing (see photo 7 on the facing page) to provide a smooth surface for snapping chalklines for the foundation walls. The smooth surface also makes it easier to sweep off the dirt that always gets on the footing while the forms are being stripped.

The final part of the pour is making a keyway, centered on top of the footing, that will lock the foundation walls in place. We usually make our keyway 1½ in. deep and 3½ in. wide, or the size of a 2x4 laid flat. We simply press a short piece of 2x4 into the concrete and drag it down the center of the footing (see photo 8 on the facing page). By

Rebar is inserted into the wet concrete. The footing forms are filled to grade, and the concrete is worked with the flat side of a shovel until the footings are level. Then ½-in. rebar is pushed down into the fresh concrete (6).

The top is troweled smooth (7), and a keyway is formed by dragging a 2x4 down the center of the footing (8).

now the concrete should have cured enough for the 2x4 to leave a significant depression.

Cutting a keyway this way causes a slight buildup of displaced concrete, and we've found that concrete built up in the corners can interfere with setting the foundation forms plumb. To avoid this problem, we end the keyway short of each corner or angle in the foundation. Because corners and angles are the strongest parts of a foundation wall, we aren't worried about compromising integrity at these points.

Before stripping the forms, we square off the foundation-line marks on the forms and mark the foundation corners in the green concrete (see photo 9 at right). A long steel bar makes short work of popping the brackets, and with the duplex nails, removing the forms is a breeze.

Rick Arnold and Mike Guertin are contributing editors to Fine Homebuilding *and the authors of* Precision Framing, *published by The Taunton Press.*

Foundation corners are marked on top. While the concrete is still green, the corner of the foundation is marked on the footing (9).

Forming and Pouring Foundations

■ BY RICK ARNOLD AND MIKE GUERTIN

The lime-green neon sign flashed "Learn your future today." Our crew, skeptics all, filed in through the purple door in the tiny storefront and listened expressionless as the gypsy inside delivered eerily detailed portraits of each of us as well as glimpses into what was in store for our lives. Rick shrugged off her bizarre prophecy that his future would involve "packaging the earth." Then, a couple of years later, he decided to buy a concrete-form business from a friend. On hearing the news, Rick's wife said, "My God, Rick, that lady was right: You are going to be packaging the earth."

There are more than a dozen different types of forming systems for poured-concrete walls, but the basic concepts of laying out, squaring, leveling, pouring, and finishing are common to any good foundation job. So whether you're pouring a foundation yourself or paying someone to do it for you, it's important to understand the process because things can go wrong in both subtle and dramatic ways.

The Foundation Hole Must Be Level

The key to raising foundation forms quickly and securely is having a good base to set them on. We always prefer to put a foundation on concrete footings as we did for the job featured in this article. The footing surface is flat and level, so the forms go up as quickly as we can move them.

The alternative to using footings is setting the forms on a gravel or crushed-stone base on the floor of the foundation hole. A good foundation hole is usually within 2 in. of level. Any more than that means a lot of labor for us scratching down or filling up to level as we set the form panels.

The excavator is also responsible for the amount of overdig, or the area that is dug beyond the actual perimeter of the foundation. Ideally, there should be 4 ft. to 5 ft. left between the forms and the hole sides to give us the room we need to handle the forms and to work on the outside of the walls. For safety's sake, the sides of the hole should not be excavated vertically or undercut.

Locate the foundation corners. Working on the floor of the foundation hole or on top of the footings, you first have to find the outside point for every corner or jog in the foundation (**1**).

Instead, there should be some sort of slope or pitch to the hole walls.

Another critical duty of the excavator is making sure that the concrete trucks have good access to the foundation (see the photo on p. 105). If they don't have access, we end up having to push the concrete by hand along the forms, sometimes 40 ft. or more. If site conditions deny even poor access, we call in a concrete pump truck to the tune of an extra $500 to $600. Once the hole is satisfactory and the footings are ready, we give the tops of the footings a final sweep before layout.

The Foundation Is Laid Out on Top of the Footings

We plot out the foundation walls the same way we did the footings. The goal is to pinpoint every corner and angle on the outside face of the foundation wall. Before we stripped the footing forms, we had scribed the corner points of the longest foundation wall on top in the concrete from points that had been marked on the forms. We double-check the distance between these points as well as the distances from the surveyor's stakes.

When we laid out the footings, we figured out all the diagonal measurements to find the location of the other corners and jogs in the foundation. Those same calculations are used again on top of the footings

Check the layout for square. Diagonal measurements are then taken to check for square (2), and chalklines are snapped for lining up the forms (3).

for the foundation (see photo 1 on the facing page). When all our points have been established, we take diagonal measurements to be sure the layout is square (see photo 2 above). We then snap lines between the points that represent the outside face of the foundation walls to give us a guide to follow as we set the forms (see photo 3 above).

Forms Are Set Up in Pairs

As a couple of crew members work on the layout, the rest distribute forms around the perimeter of the hole, sliding them down along the sides of the hole in pairs. By the time the layout is done, enough forms are in the hole, and the layout crew can begin setting them up.

The forming starts in one corner. We use 90 degree forms for inside corners, but for outside corners we have special brackets that join two standard forms at a right angle (see photo 4 on p. 108). Once the corner forms are set up, we tweak them until they're exactly plumb, using shims if necessary. Plumbing each corner precisely simplifies the later job of squaring the top of the foundation.

When a corner has been set, crew members each take a direction and begin setting up the standard-size (2-ft.-wide) forms. To join our forms together, metal Ts are slipped through reinforced holes in the side rails of the form that is already set up. Then flat, slotted foundation rods are slipped over the ends of the Ts, locking the inside and outside panels at the specified width for the foundation wall, in this case 10 in. (see photo 5 on p. 109). Foundation rods not only de-

TIP

If a wall length is specified to an odd inch or to a fraction, nail a spacer made of ¾-in.-thick furring or the appropriate-size plywood to the ends of an inside and an outside panel to make up the difference.

Steel Reinforcement

Incorporating steel reinforcement, or rebar, into concrete walls significantly increases the foundation's ability to resist the lateral force from the soil around it and helps control the cracking that occurs in many walls.

Some regions, most typically in seismic zones, require rebar be included in the foundation. In others, it's required by common sense. In residential foundations ½-in.- or ⅝-in.- diameter rebar is used and the grade of steel is either 40 or 60, which is stamped on every piece.

Exact rebar placement and pattern is determined by the local building codes, blueprints, engineered drawings, or, if not required by any of these, it is placed according to the building contractor's instructions.

Some designs call for both horizontal and vertical placement, creating a grid of steel buried in the wall. Others require only horizontal rebar. Although not shown, the foundation in this article received a double row of ½-in. rebar about a foot from the bottom and a double row about 9 in. from the top—typical for our region.

Forming starts in a corner that uses a right-angle form for the inside and special brackets for the outside (4).

termine the width of the wall but also—along with the Ts—keep the forms from spreading during the pour.

The next panels, one inside and one out, are placed close enough to the previous ones to feed the ends of the Ts into the matching holes. Then the forms are pushed hard against each other, and a flat, tapered pin, or wedge, is slid into the slot on the end of the T (see photo 6 below), locking the two panels together. At this point, the wedges are left loose in the slot until the walls have been squared and straightened.

Our forms come in 2-in. wide increments from 2 in. to 24 in., so it's easy to anticipate what we'll need for each wall. A 30-ft. 2-in. wall will use thirty 24-in. panels plus a pair of 2-in. fillers. If a wall is 41 ft. long, we use forty 24-in. panels and a pair of 12-in. panels. If a wall length is specified to an odd inch or to a fraction (we just love those), we simply nail a spacer made of ¾-in.-thick furring or the appropriate-size plywood to the ends of an inside and an outside panel to make up the difference. The panels with spacers are then locked to neighboring panels with longer Ts.

Besides 90-degree corners, the most common angle we're asked to form is 45 degrees. We have forms and brackets similar to our 90-degree system to create 45-degree angled walls. However, for other angles, we usually build forms on site.

The house featured here had sections offset from the main body of the house at 30 degrees and 60 degrees. We formed these angles by locking smaller-size panels together with perforated-steel strapping. The inside panels are connected to the outside panels with ¼-in. steel rod (called pencil rod). We bolt clamps onto the pencil rod to hold the panels at the 10-in. wall width. Voids between the forms are filled with rigid-foam insulation that we cut and insert.

Slotted rods tie inside forms to outside forms. Slotted rods that slip over Ts hold the forms apart at the right wall width (5).

Tapered pins, or wedges, through the Ts hold adjacent panels together (6).

Bracing keeps the walls plumb during the setup and pour (7).

Strings help the crew to straighten the walls. Before the concrete trucks arrive, strings are stretched as guides for straightening the walls (8).

Strings Straighten and Square Walls

As the forms go up, we brace both the inside and outside sections with 2x bracing every 8 ft. to 10 ft. (see photo 7 at left) to keep forms from racking and falling over. When all the panels are up and the job is closed in, we slip steel channel over the tops of the forms. The channel fits tightly, locking the panels into alignment with each other.

Next we string the whole job to straighten and square the foundation (see photo 8 below). Mason's twine is stretched from one corner to the other; the string is kept lined up exactly with the face of the outside corner panels. Then a crew member walks along the top of the forms, telling two other crew members who are on the ground (one inside, one out) which way to adjust the 2x braces to straighten the wall. Braces are not nailed; instead, they are wedged under a horizontal member of the form.

The next job is squaring the foundation by measuring diagonally between the corners just as we did when laying out the foundation on the footings. If need be, we shift a corner or two until our diagonal measurements are equal. When the entire job is squared, we once more straighten the tops of the walls and tighten the bracing.

Next we check and tighten all the hardware, the easiest but probably most crucial job of the process. Each foundation rod is checked methodically inside and outside to make sure that it is properly engaged by a T. If a T misses the slot in the rod, it can cause the wall to blow out from the weight and pressure of wet concrete. As the rods are checked, each wedge is driven home to tighten the joints between the forms.

The top of the foundation is found with a transit (9), and chalklines are snapped at that level (10).

A Transit Helps to Find the Top of the Pour

We're now ready to shoot the grade, or the height of the foundation wall (see photo 9 above). Because we're on footings, we shouldn't have any high or low spots to factor in. Unless otherwise specified, the grade is established at 93 in. from the bottom of the form, and a 6d grade nail is driven at that elevation. (Our forms are 96 in. tall; our alignment channels are 2½ in. deep. A pour height of 93 in. keeps the concrete off the channel.) A yardstick is then placed on top of that grade, and the height is read through our transit.

The crew member with the yardstick works his way around the foundation, set-

Basement-window frames are inserted and tacked to the forms (11), and form-release oil is applied to the inside of the forms (12).

ting grade nails at each corner on the outside form at the height indicated by the crew member at the transit. If a wall is longer than 24 ft., we set a grade nail in the middle of the wall. Lines are snapped between the grade nails (see photo 10 on p. 111), and then extra nails are set in the snapped line about every 4 ft.

Before the channel went on, we'd slipped the basement-window frames down between the forms (see photo 11at left). At this point, the frames are brought up to the grade line and tacked to the inside of the forms. We also locate and nail in metal forms for the beam pockets. These forms leave a small shelf inside the wall to hold the end of the main beam. If the foundation requires a sewer chase or a chase for any other purpose, we slide a Styrofoam® block between the forms and secure it at the proper location with 16d duplex nails driven from the outside. These blocks are made on site to the required dimension and fit tightly between the forms.

The inside faces of the forms have to be coated with form-release oil to prevent the forms from sticking to the concrete as it cures. We usually apply the oil just before or as the trucks arrive (see photo 12 at left). There are many types of form oil and release agents. Motor oil and diesel fuel used in the old days have been replaced with more environmentally friendly nontoxic mixtures. Many places require the use of these newer products, and we've had good luck with paraffin and vegetable-oil-based release agents. We use a handheld pump sprayer with a wand to apply a light coat on the panel faces.

Concrete Has to Be the Right Mix

When we order concrete, we specify the strength of the mix and the slump, or stiffness of the mix. A 2,500-psi mix provides sufficient strength for most house foundations, and most mixes arrive at a slump of

Grade nail

When the prep work is done, the forms are filled with concrete. The backside of a shovel floats the top of the concrete until it's halfway through the grade nails (13). The top is then screeded with a 2x4 to smooth it out (14).

15

Anchor bolts are inserted while the concrete is still soft (15).

stayed straight, adjusting the forms if necessary. Our concrete mix is generally loose enough that we don't have a problem with voids from pockets of trapped air. But if a real stiff mix is specified for a job, we tap the forms with a rubber mallet, and in extreme cases we use a vibrator to remove trapped air from the mix.

With the forms filled, we grade the top surface by floating the back side of a shovel on the concrete and working it to level the surface (see photo 13 on p. 113). At this point, we also add or remove a touch of concrete until half of each grade nail is exposed.

A crew member then follows along, screeding the surface with a length of 2x4 (see photo 14 on p. 113). The edge of the 2x4 is pushed vigorously up and down, forward and back, to bring water up to the surface, making it smooth and level. The final step is inserting ½-in. anchor bolts into the top of the wet concrete (see photo 15 at left). Some local codes may require that the bolts be in position before the concrete is poured to ensure proper aggregate consolidation around the bolts. But our building officials allow us to insert the bolts directly into wet concrete, moving them up and down slightly to make sure there is no trapped air around them. When we poured this job, CABO code required bolts every 6 ft. Here in Rhode Island, that requirement has since been upgraded to anchor bolts every 4 ft.

Stripping Forms: Set Up in Reverse

The concrete is allowed to set up overnight, and the next morning, we're back at the site to strip the forms. Stripping is basically the setup process in reverse. First, we remove the string and the channel from the tops of the forms. Then the Ts, wedges, and any other removable hardware come off.

If the form oil has done its job, the panels should pop away from the walls easily (see photo 16 on the facing page). The excess

TIP

Because concrete starts to cure within two hours, begin the pour with the most time-consuming or hard-to-reach areas of the foundation and work your way around the foundation from there.

around 4, a medium number that flows readily. Occasionally, we ask the drivers to add a little water and spin the load if it seems too stiff. However, too much water can weaken the concrete, so we're careful to add only a small amount to make the concrete easier to manage without diminishing its strength.

Because concrete starts to cure within two hours, we begin the pour with the most time-consuming or hard-to-reach areas of the foundation and work our way around the foundation from there. This strategy keeps any of the walls from curing before the other walls can be poured.

During the pour, we keep a constant watch for blowouts or shifting panels. We stop the pour as needed to rebrace walls, keeping the forms aligned and plumb. When the concrete nears grade, we slow the pour rate a little and move the chutes to keep from overfilling the forms.

Once the walls are topped off, we check the strings again to make sure the walls

Forms are stripped and carried to the truck. After the concrete has cured overnight, the crew strips off the forms (16) and takes them to the truck to be stacked.

As each form is removed, excess concrete along the top is scraped off (17).

concrete that built up at the grade line is scraped off the forms (see photo 17 above right), which are brought to the truck.

After the forms have been stripped, we snap off the ends of all the foundation rods that held the forms together and that now project beyond the foundation (see photo 18 at right). The rods are scored so that they break off safely below the surface of the foundation wall. We remove the rods to make the job site safer for the rest of the subcontractors and to allow the exterior of the foundation walls to be damp-proofed.

It's best to let the concrete cure before backfilling the foundation. Concrete doesn't reach its full compressive strength for 28 days or so, but 5 days to 7 days is usually sufficient curing time for the concrete walls to withstand backfill pressure. Still, it's always a good precaution to brace the inside of any green walls that are long and straight before backfilling against them.

The ends of the foundation rods are then broken off with a hammer (18).

Rick Arnold and Mike Guertin are contributing editors to Fine Homebuilding *and the authors of* Precision Framing, *published by The Taunton Press.*

Footing and Foundation in One Pour

■ BY JIM BLODGETT

We have been pouring foundations pretty much the same way that our grandfathers taught us: Build the forms for the footings; pour the concrete; strip the forms. Build the forms for the foundation; pour the concrete; strip the forms again.

This time-tested approach to footing and foundation construction works fine for big jobs, but it's not the best way to tackle a small job. You pay a pretty stiff upcharge when you order a short load of concrete (less than 4 cu. yd.), and small projects usually call for short loads. Pouring the footing and foundation at the same time can convert two short loads into one regular load.

When pouring footings and foundation walls separately, there's a built-in safety net: If the footings aren't perfectly square, you can correct mistakes when you form the walls. Not so with a monolithic pour; you need to be more careful when laying out and setting up the forms.

Batterboards guide stringlines, which in turn guide the placement of the footings. Specialized metal straps and brackets position the wall forms on the footings.

Build Forms to Disassemble Easily

Another important thing to remember when building forms is that you will have to strip them. Poorly thought-out joints can make stripping forms a nightmare.

Forms with outside corners generally are easy to strip because pressure from the concrete tends to push the joints apart. Inside corners, however, are pushed together. When building forms, I follow a couple of simple rules for inside corners. If a formboard butts existing concrete (the house's foundation), I butt the opposite end into wood by reversing the lap of the corner, and I use two short boards joined with a plywood bracket instead of one long board.

Buggy-pouring is easier with a helper. A strategically placed helper with a splash board gets the mud inside the forms with fewer spills. If you can't persuade your daughter to help you, brace and nail the splash board in place.

Clean the forms while the concrete is wet. Stripping goes more quickly if you don't have to chip away hardened concrete to find the nails. Another way to make stripping easy is to use two short boards rather than one long board when building the forms. This reduces the effect of pressure exerted by concrete on the inside-corner joints.

Keep Concrete Where It Belongs: In the Forms

If placing forms accurately is one difficulty encountered in a monolithic pour, keeping the concrete in the forms is another.

On small jobs, space is usually tight. Getting concrete to the forms may require a pumper or a wheelbarrow. But even distributing concrete within the forms can be tricky on a tight lot. If you don't have room to dump the wheelbarrow every couple of feet along the forms, an electric vibrator can be helpful. I rent an electric vibrator to move concrete within the forms and to reduce air and rock pockets (honeycombing). The vibrator is a wonderful tool, but it has drawbacks; specifically, vibration adds pressure to the

Straps and Brackets Support the Wall Forms

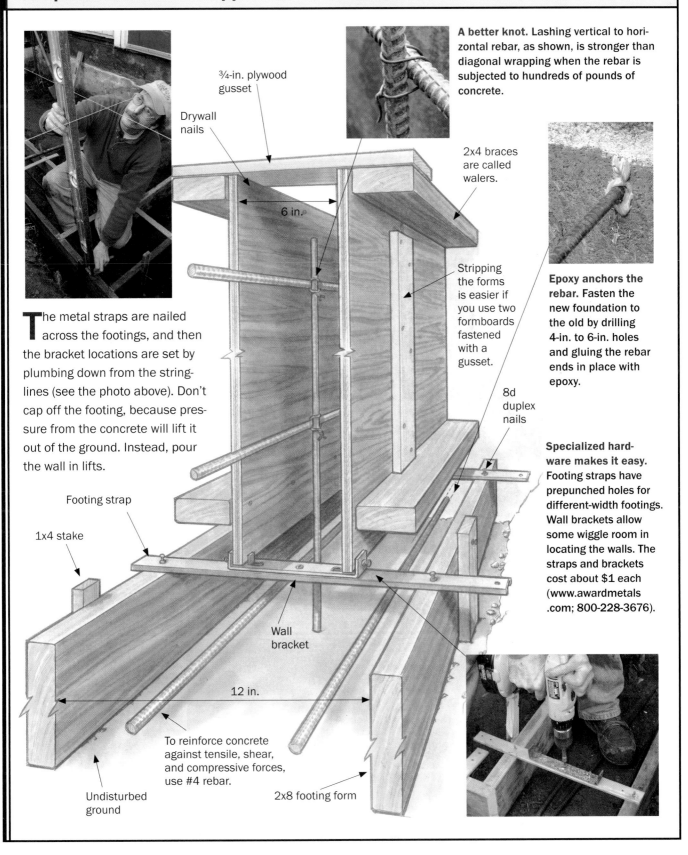

A better knot. Lashing vertical to horizontal rebar, as shown, is stronger than diagonal wrapping when the rebar is subjected to hundreds of pounds of concrete.

¾-in. plywood gusset

Drywall nails

6 in.

2x4 braces are called walers.

Stripping the forms is easier if you use two formboards fastened with a gusset.

Epoxy anchors the rebar. Fasten the new foundation to the old by drilling 4-in. to 6-in. holes and gluing the rebar ends in place with epoxy.

The metal straps are nailed across the footings, and then the bracket locations are set by plumbing down from the stringlines (see the photo above). Don't cap off the footing, because pressure from the concrete will lift it out of the ground. Instead, pour the wall in lifts.

8d duplex nails

Specialized hardware makes it easy. Footing straps have prepunched holes for different-width footings. Wall brackets allow some wiggle room in locating the walls. The straps and brackets cost about $1 each (www.awardmetals .com; 800-228-3676).

Footing strap

1x4 stake

Wall bracket

12 in.

To reinforce concrete against tensile, shear, and compressive forces, use #4 rebar.

Undisturbed ground

2x8 footing form

Batterboards and Strings Align the Forms

Locating the forms for this L-shaped foundation wall can be done by stringing lines between batterboards and the walls of the house.

Set the long string by measuring equally off the house at two points. Then set the shorter string square to the house by applying the 3-4-5 triangle rule. A triangle with these side lengths contains a 90-degree corner. Because the strings are for reference, they don't have to be at any particular height, but they should be level.

Some people establish a level grade line inside the foundation forms and pour to the line, but setting the top of the form to the finished height of the wall makes the whole process easier.

Long string is parallel to house.

Batterboards

forms, notably at the bottom and on corners, which can cause the forms to break. Anyone who has poured concrete has learned about this hydrostatic pressure the hard way.

In monolithic pours, you have the additional challenge of controlling a lavalike mudflow from the open top of the footing. You can't contain the concrete by sealing this opening, so don't try. The same hydraulic pressure that lifts concrete over the top of the form can lift sealed forms out of the ground—stakes and all—leaving you with a monolithic mess.

The key step is keeping an eye on the opening while you pour and vibrate. If concrete starts to ooze over the top of the footing, move to another spot and let the concrete stiffen before you add more weight on top of it. When you come back to it, vibrate the two pours together enough to ensure a good joint, but not enough to lift the forms.

Jim Blodgett lives in Roy, Washington. His website is www.asmallwoodworkingcompany.com.

Tight Corners Are Easy to Bend

Using two lengths of black pipe, the author gets excellent results without an expensive rebar bender.

Put one pipe's opening at a corner.

Slide the second pipe over.

Bend to the desired angle.

Voilà!

A Vibrator Strengthens and Smoothes Concrete

To get a smooth finish on a concrete wall, I vibrate the concrete as I pour it. Vibration also adds strength because it removes air pockets and eliminates the cold joints between successive pours. I use a series of plunges into the concrete instead of constant vibration, which would create too much pressure on the forms and make the concrete soupy. At $40 a day, a ¾-in.-dia. vibrator is worth renting.

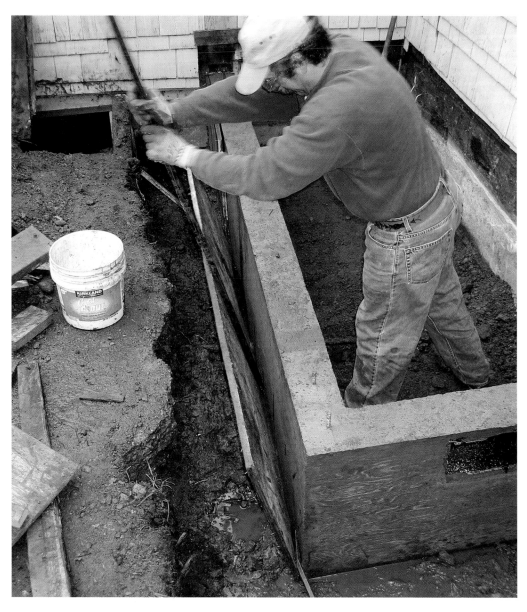

Outside corners strip easily. The pressure of the concrete doesn't act to tighten the joints in the formboards on outside corners as it does on inside corners. When stripping outside corners, the author starts at one end and works his way to the other. Inside corners should be formed so that a central section can be pulled away and stripping can proceed in both directions.

Insulated Concrete Forms

■ BY ANDY ENGEL

On a Saturday morning two summers ago, helped by my wife, Pat, and four usually deskbound coworkers, I poured the concrete walls for my new home's 28-ft. by 40-ft. basement. Containing the heavy slurry of concrete were insulating concrete forms (ICFs) made of foam and plastic ties (see the photo on the facing page).

Setting the forms, braces, and rebar took Pat and me about two weeks of part-time work. However, fretting about pouring three truckloads of concrete into forms that resembled a hybrid of take-out coffee cups and Lego® blocks cost me a lot of sleep during those weeks. When full, every 5 ft. of those forms would hold 1 cu. yd. of wet concrete and weigh about 2 tons. The potential for the forms to burst, topple over, or simply sneak out of plumb was real. All went well, though, and by early afternoon, the crew was sitting in the shade, drinking cold beer and grilling burgers.

I could have hired a concrete contractor and been done with my basement in days, for about the same money. But I wouldn't have gotten the insulated basement that I wanted. If the thermal mass and the air-sealing effects of the concrete are considered, industry sources claim effective R-values that push 40. I don't know if that number will stand scrutiny, but I know that from the foam alone, my basement walls have a respectable R-value of 25.

Also, I poured those basement walls during a building boom. Contractors in my area were scarce. With care, ICF walls can be built by anyone with carpentry and concrete experience, so I avoided the headaches of finding one more contractor.

If I ever build another house, I'll consider taking ICFs all the way to the roof. The cost of ICF walls is somewhat higher than that of frame construction, but this factor is less troublesome in areas where wind and seismic loads are considerable. In these areas, houses built of steel-reinforced concrete—that are well insulated to boot—make sense.

One job does not make me an expert, but it did make me curious. So I spoke with experienced ICF builders from Maine to Florida to Washington. There are more than 40 ICF manufacturers, and their systems assemble in an almost-equal variety of ways (see the sidebar on pp. 128–129). I don't have space to explore their proprietary ins and outs. The differences mainly have to do with how the blocks attach to each other, not with how you set windows and doors, or how you brace and fill forms with concrete. Talk to your distributor for brand specifics.

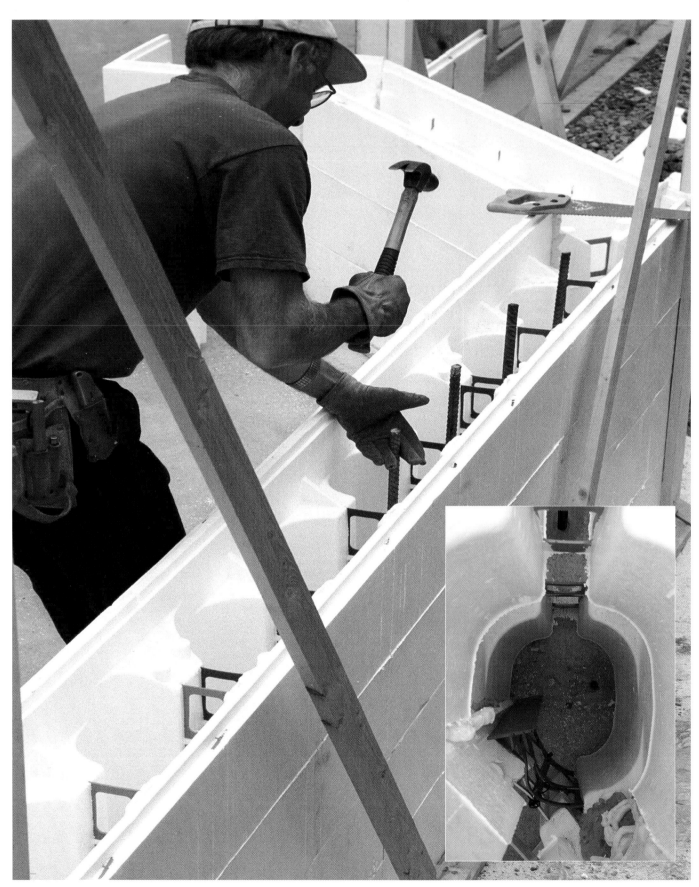

Hammering rebar into drilled holes (inset) ties wall to footing. Other methods include a keyway in the footing, L-shaped rebar cast in place, and rebar epoxied into drilled holes.

It's Easiest to Start on Level Footings

Most of the builders I spoke with become concerned when the footings are ¼ in. or more out of level. All of them fix the problem with the first ICF course.

If the footing is humped, determining its height with a laser level or a transit is simple, and you can shave the bottoms of the forms to fit with a saw or a Surform® plane.

Depressed footings are easily dealt with by finding the high point and shimming the ICFs level with it. Some contractors shim with mortar; others use foam scraps held in place with gunnable low-expanding foam such as Pur-Fil from Todol Products Inc.

If footings need to be stepped, Chris McCormack, an ICF contractor from Hartford, Connecticut, suggests doing so in increments equal to the ICF height, thereby avoiding long horizontal cuts.

Tying Walls to Footings

Requirements for tying walls to footings vary geographically. Here in seismically lazy Connecticut, I grooved a keyway into the fresh concrete of my footings with a 2x4 scrap. In the process, I created a hump in my footings; check yours for level after grooving.

In many cases, specs call for rebar dowels to link the walls and footings. L-shaped rebar dowels cast into the footings make the strongest bond.

Where seismic or wind loads are of little concern, builders frequently set the first course of ICF blocks, then drill holes 2 in. or so into the footings for the rebar (see the photo on the facing page). If done within a few days of the pour, the drilling goes fast. And accurate layout is easier, particularly with grid systems whose rebar must be centered in the form's vertical channels. Some

builders epoxy rebar in place. Others simply hammer it home.

Drilling in the rebar offers a margin of job-site safety. Rebar sticking out of a footing can turn a stumble into a goring. Placing a course of forms first at least offers a visual clue to the rebar's presence.

Typically, the dowels should protrude from the footing by at least 40 diameters. This placement ensures a solid bond with the vertical rebar in the wall. So #5 rebar, whose diameter is ⅝ in., should stick out of the footing 25 in. To be sure that the wall's vertical rebar comes close enough to the dowels to establish structural connections, some builders place 2-in. sleeves of 1½-in. plastic pipe around the dowels. Once all the forms are set, the vertical rebar sets into the pipe sleeves and is wired to the horizontal.

Cleats Guide the First Course of Forms

The next step is snapping perimeter chalk-lines on the footing. ICF users typically snap these lines to the inside of the walls because they set the forms from inside. They then nail a cleat alongside this line to guide the first course. After setting the first course of forms, most builders nail another cleat outside the forms. The cleats reinforce the form's bottom edges, where the concrete pour generates the greatest pressure (see the photo on p. 126).

I've seen cleats made of 1x3 strapping and of 2x4. Todd LaBarge builds with ICFs on Cape Cod, Massachusetts. He nails down 2½-in. steel studs, channel up, inside the chalkline. His forms' 2½-in.-thick sides firmly lock into the studs. The studs stay in place, saving him the labor of stripping the cleats.

Bill St. Laurent, a builder in Kittery Point, Maine, places a cleat outside the walls. He sets the first ICF course, then has the basement slab poured. Pinched between the

TIP

The dowels should protrude from the footing by at least 40 diameters. This placement ensures a solid bond with the vertical rebar in the wall.

A cleat holds the bottom course in place. It also reinforces the forms in the area subject to the greatest pressure during the pour.

Cutting Weakens ICFs

Some ICF form systems have ready-made corners; others must be cut and joined on site. In either case, the outside of the corner usually doesn't have built-in ties, as does the rest of the form. It's a weak spot that most ICF builders routinely brace.

Corner braces can do double duty. Stoutly braced and plumb, they not only reinforce the forms during the pour but also make a reliable starting point for the succeeding courses of forms (see the top left photo on the facing page).

I made my corner braces by nailing two 2x10s into an L-shape. That decision was overly cautious. St. Laurent builds his braces from 1x4s that are primed to minimize the chances of warping. Jeff Preble, an ICF manufacturer and contractor in Gray, Maine, uses 4-in. aluminum angles that he carts between jobs, and Juen uses almost any kind of lumber that he can later reuse in the house.

Most ICF builders work inward from the corners, minimizing cuts by placing them at window or door openings. (The favored tool for cutting is a cordless circular saw.) Cuts can weaken the forms. Plastic, steel, or foam webs tie together the form sides. Each web reinforces the form for half the span to the next web. Cutting between the webs, particularly if more than half the span remains, reduces a form's ability to withstand the pressure of wet concrete (see the top right photo on the facing page).

When stacking ICFs, Jeff Preble cautions to be sure that the webs stack atop each other. They are where finishes—drywall, siding, and the like—attach. Changing that layout halfway up a wall guarantees an angry drywaller. Preble further warns to line up the vertical channels of grid or post-and-beam ICFs (see the bottom right photo on the facing page). Doing otherwise can compromise the wall's final strength.

> *Corner braces can do double duty— they reinforce the forms during the pour and also make a reliable starting point for the succeeding courses of forms.*

cleat and the slab, his forms stay put, and the slab gives St. Laurent a clean, level working surface. Cleating to reinforce the forms may be overkill, however. Neither Peter Juen of Miami, Florida, nor Matt Shacklford of Mississippi—both ICF users—installs cleats; they simply glue the forms down with low-expanding foam. To speed layout, Juen lays out door and window openings on the slab before setting any forms.

Branford, Connecticut's Jim Eggert sets the first course of forms in the houses he builds before the footing sets. Bonded to the concrete, the forms need no guide boards.

Corner braces support a relatively weak spot in the forms. Setting the braces dead plumb provides a good starting point for stacking forms.

Bucks Create Window and Door Openings

To dam the concrete at openings and to create an attachment point for windows and doors, ICF users usually build bucks from pressure-treated 2x stock (see the photos on p. 130). The inside of the buck forms the rough opening for the window or door.

The buck stock is ripped, if need be, to the overall width of the forms. Two 2xs on the flat, sized to leave a 3-in. gap in the middle through which the concrete can be placed, comprise the bottom of the buck. Each is nailed so that its outside is flush with the buck's legs.

Potential Trouble Spots

Cut blocks are the weakest point. Cut blocks should include a web and be joined to the adjacent blocks with low-expanding foam (top photo). Plywood scabs provide further reinforcement. The bottom photo shows a misaligned cavity in a grid ICF; it reduces structural integrity and throws fastening surfaces out of sequence.

Most builders screw or nail flanges of 1x4 or plywood strips around the buck's perimeter. These strips capture the ends of the forms and reinforce cut forms. My ICFs

Choosing ICFs

There isn't space here to describe each of the 40 or so systems on the market. But generally, there are three main types of forms. The most common are hollow blocks, whose 2-in.-thick sides are held together with plastic or steel connectors that extend to the faces of the block, also providing a fastening point for wall finishes. These blocks are typically 16 in. high by 48 in. long. Their overall width depends on the concrete thickness, which can vary from 4 in. for an above-grade wall to 12 in. or more for basement walls.

Plank forms are similar to blocks except that the foam sides are longer and narrower, typically 1 ft. by 8 ft. Panel forms, the third type of ICF, have sides as large as

4 ft. by 8 ft. Additionally, the three form types shape the concrete within in one of three ways (see the drawings on the facing page). The simplest form flat concrete walls. Apart from the integral foam insulation, these walls are no different than traditionally formed concrete walls. Grid ICFs form walls whose concrete, if the foam were stripped away, would resemble giant waffles. In the thinnest spots, the concrete may be only 2 in. thick, and in the thickest, it will usually be 6 in. or 8 in. Grids use less concrete than do flat walls, but because the concrete can hang up in thinner areas, they are more susceptible to voids. The third configuration is post and beam. As the name suggests, the concrete in these walls

ICFs Come in Three Main Types

BLOCK FORMS

Block forms commonly are 16 in. by 4 ft.

PLANK FORMS

Plank forms are typically 1 ft. by 8 ft.

PANEL FORMS

Panel forms come in sizes up to 4 ft. by 8 ft.

is formed into one or more beams supported by posts. The foam of the forms is all that fills the spaces within this concrete armature. Because of this fact, these systems may be unsuitable for below-grade use where they must resist a considerable amount of backfill pressure.

Grid-wall and flat-wall block systems are the most common combinations, and all call for reinforcing steel.

Additionally, there are several ICFs made from cement composites. The units are heavier than foam ICFs. They offer several advantages, though. Concerns have been raised that carpenter ants and termites can burrow unseen through below-grade foam and into houses. Some building codes, particularly in the South, require below-grade foam to be treated to resist these pests. Good detailing should sidestep the problem in much of the country regardless. But composite ICFs are one way that you can be more sure.

Additionally, the entire surface of most composite ICFs accepts fasteners, compared with foam ICFs, whose fastening strips are on 8-in. or 12-in. centers. No interior drywall is needed either; there is no smoke-generating foam to cover, and a coat of plaster will finish these walls nicely.

ICFs Form the Concrete in Three Ways

FLAT WALL   **GRID WALL**   **POST-AND-BEAM WALL**

Flat walls are essentially similar to traditionally formed walls.

Grid walls use less concrete than flat walls.

Post-and-beam walls use the least concrete.

were 13 in. wide, wider than the available lumber, so flanges weren't an option. I centered the bucks on the ICFs and held them in place with steel pipe straps screwed to the webs.

Bracing bucks square is critical. Because bucks must resist the weight of the concrete without bowing, most builders also brace between the bucks' sides with 2x stock. Wide openings with concrete poured above them get a few 2x posts. Pouring concrete above an opening creates a cast-in-place lintel. Rebar size and placement are critical here, especially with wide openings or heavy loads above. This place is one where an engineer can be well worth the cost.

Typically, nails, screws, or lag bolts are driven into the back of the buck (but not so far that they stick out of the front) before it's installed. Concrete hardens around these fasteners, locking in the buck.

ICFs Need to Be Braced

When the forms reach 5 ft. or so in height, most builders install vertical bracing inside the wall, usually 2x4s or 20-ga. steel studs on about 6-ft. centers. Both sides of windows, doors, and corners should be braced. These vertical braces are screwed to the webs, and screwed or nailed to the guide cleats on the footing. A diagonal 2x4 kicker will fasten to the vertical brace and be staked to the ground.

Setting the upper forms and pouring the concrete is easiest from a scaffold. One-time ICF builders often make a scaffold support with a 2x4 top and plywood sides that fasten to the vertical brace. The kicker slides between and is fastened to the plywood sides. Turnbuckles at the bottom of the kicker, commonly used on traditional concrete forms and available at most masonry-supply yards, push or pull the wall plumb.

Similar rigs are available from the E.M.M. Group, Easy-Wall Bracing & Alignment Systems, and ReechCraft Inc. AAB Building Sys-

Diagonal braces hold pressure-treated window bucks square. Flanges capture and reinforce cut-to-fit ICFs. Wood strips on the legs will key into the concrete. Kickers and turnbuckles (right) hold bucks plumb for the pour.

A homemade hole saw neatly inlets utility sleeves. A few saw kerfs in the end of a pipe make a quick and accurate drilling tool.

tems, an ICF manufacturer, rents a scaffold and bracing system with its forms.

Placing Rebar

Installing utility penetrations is easiest before placing rebar. Commonly, builders install a sleeve the next size up from whatever size pipe is called for. To make a perfectly sized core drill, make a series of saw kerfs in the end of a piece of sleeve material that's about 1 ft. longer than the wall is thick (see the photo above). I braced the forms around the sleeves with 12-in. squares of ¾-in. plywood whose centers were cut out to fit around the sleeve. The universal fastener, 1⅝-in. drywall screws, held the plywood to the webs.

Every ICF contractor I asked installed the horizontal rebar as the forms rose. The webs holding together the forms also support the rebar, and it's a simple matter to lay it in place as you go (see the photo at right). My ICFs called for one horizontal rebar every course. Rebar placement is critical, and you should look to your ICF manufacturer or engineer for sizing and spacing guidance.

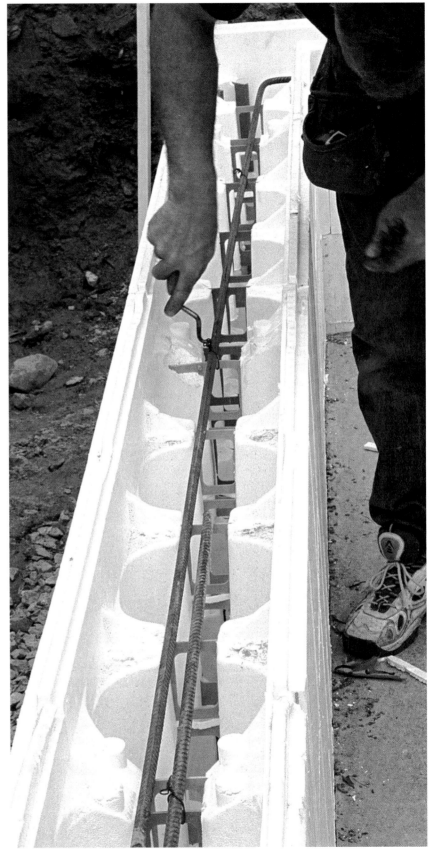

Horizontal rebar is placed in block and plank ICFs as each course stacks. With panel systems, horizontal rebar slides in from the wall ends. With all systems, vertical rebar is placed from above after the top course is set.

In one variation, steel joists are set and encased in concrete (top). Supported below on a ledger, the floor is sheathed and provides a work platform for pouring walls. In another variation (bottom), bolts support a ledger. Forms are cut to bring concrete to the ledger, providing a surface for bolts to pinch the ledger against.

Forming to Support a Floor

Because I built only my basement from ICFs, I treated the top of the wall as any foundation. After the pour, we smoothed the concrete and inserted anchor bolts. Todd LaBarge cuts the mudsills to length, drills them, and then inserts the anchor bolts, washers, and nuts while the sills are on the sawhorses. Before the concrete sets, he snaps chalklines and sets the sills.

Bill St. Laurent uses no mudsill, instead setting steel joists directly in the concrete (see the top photo at left). He braces forms from the outside, rather than the inside, using turnbuckles and 2x4 kickers (see the bottom photo at left). Because the slab inside is already poured, moving a stepladder or rolling scaffold to set the upper forms is easy. After straightening the walls, St. Laurent sets the joists, flush with the wall's top and about halfway in, through slots cut in the top form. A 2x4 ledger screwed to the webs and supported by vertical 2x4 posts props up the joists. He sheathes the floor, making a perfect working platform for the pour. St. Laurent says that this process insulates the floor's edge and also eliminates air leaks common with traditional mudsills and rim joists. His system works with either a conventionally framed house or one built entirely of ICFs.

Jim Eggert takes a similar tack, instead using wood I-joists whose ends he wraps with plastic to protect them from the concrete's moisture. Matt Shacklford sometimes uses a narrower form when doing his top course, forming a ledge where the joists will rest.

Affixing a pressure-treated ledger to the ICFs with anchor bolts and fastening the joists with joist hangers is another way to install a floor. For the ledger to attach solidly, though, the ledger must contact the concrete, at least around the bolts (see the bottom photo at left).

String checks walls for straightness. Tweaking the braces will bring the wall straight. If scaffolds brace your walls, your weight on them can move the walls, so stay off them while checking for straight.

At least one manufacturer, Southeastern Metals Manufacturing Co. Inc., makes joist hangers that set directly into the concrete of an ICF wall. If you plan to build additional ICF walls atop the first, you'll need to set rebar dowels just as in the footing.

Straighten the Walls, and Double-Check Everything

Everyone I asked checks the top of their forms for straightness with string (see the photo above). While most builders aim to get the walls straight and plumb at this point, a few lean them slightly toward the braces. Braces exert more force pushing than pulling, so if the wall goes out during the pour, a quick spin of the turnbuckle will put it right. Most builders recheck their walls immediately after they're done pouring.

Foam is lighter than concrete, and top courses have been known to float during the pour. Vertical braces usually prevent this,

but to be sure, Todd LaBarge glues down the top two courses with foam. I held mine together with filamented strapping tape.

With conscientious installation, blowouts are rare. Even so, it's a good idea to have a patch on hand. Two 2-ft. squares of ¾-in. plywood with holes in their centers, some ½-in. threaded rod, and two nuts will handle most blowouts. Blowouts occur when a form, weakened by cutting, damage, or defect, lets go between two webs. When blowouts happen, move the pour to a different part of the wall and shovel up a wheelbarrow or so of concrete. Patch the foam back in, push the threaded rod through the forms, and place the plywood on the rod. Tighten the nuts, and you're set. Many builders skip the threaded rod and screw the patch to adjacent webs.

ICFs usually bulge before letting go. Caught bulging, the fix is easy. Move the pour, and screw a piece of plywood to the form's webs over the bulge.

> *Braces exert more force pushing than pulling, so if the wall goes out during the pour, a quick spin of the turnbuckle will put it right.*

Let the truck do the work. Boom pumps such as this one don't come cheap, but you won't have to lug a concrete-filled hose. They can quickly place concrete overhead, and some have a reach of more than 100 ft.

Ordering the Right Concrete Mix Is Critical

Concrete must flow well to fill the forms, but not be so liquid as to increase the hydraulic pressure dangerously within the forms. A 3,000-psi mix at a slump of 5 or 6 is about right, although some builders get a 3,500-psi mix. Most use smaller ⅜-in. aggregate because it is less likely to hang up on the rebar or the webs and is easily pumped.

Jeff Preble has his supplier add a midrange water reducer to the mix. Water reducers make concrete flow better, without the loss of strength that would happen if water were added to improve flow. Preble also worked with his supplier to develop a custom mix that uses more sand than coarse aggregate. Additional portland cement in the mix makes up for the reduction in strength caused by reversing the usual sand-to-aggregate mixture. The mix flows so well that Preble doesn't vibrate or even tap on the forms to consolidate the concrete (more on consolidation later).

One up and one down. Having a helper at the bottom of the wall keeps the person directing the concrete apprised of problems.

A Concrete Pump Is Well Worth the Money

For speed and controlled placement, concrete pumps are hard to beat. Still, they're an expense that can be avoided on basement jobs that have good access for trucks, which is how I poured mine. Two things eased my pour. The first was a simple plywood funnel that provided a bigger target for the concrete, greatly reducing spills. The second was front-discharging trucks. Because the driver can see to place the chute precisely without ever leaving the cab, these trucks are great time-savers.

You may not have good site access, or you may be pouring walls higher than the chute of a concrete truck. Then you must pump.

There are two types of concrete pump. The easiest to work with—and the most expensive—is a boom pump (see the photo on the facing page). Some booms reach more than 100 ft., so the truck can park almost anywhere on site and reach where it's needed. The ready-mix truck fills a hopper, and the pump pushes out the concrete through a 4-in. hose.

Boom trucks generate enough pressure to damage ICFs, so have the pump supplier provide a reducer that narrows the hose end to 3 in. or 2 in. To slow the concrete further, ask for two 90-degree elbows to configure the end of the hose as an S.

Line pumps (see the photo above) are less aggressive than boom pumps. They usually come with a 3-in. hose that reduces further to 2 in. Their disadvantage comes with man-handling a 3-in. hose that's full of concrete.

Wet concrete exerts tremendous hydraulic pressure. Aim the concrete at webs to break its fall, and pour in lifts of perhaps 4 ft. It's also important to fill both sides of corners and bucks evenly.

While I watched, Jeff Preble spent three hours pouring a foundation using a line pump. In that time, 66 tons of concrete passed through his hands.

Filling the Forms with Concrete

No matter how you transport the concrete from the trucks to the forms, wet concrete exerts hydraulic pressure that can damage ICFs. All the contractors I spoke with aim the concrete at webs to break its fall (see the photo above).

All the manufacturers' manuals that I read and most of the contractors I spoke with stress pouring the concrete in lifts (or layers) of 2 ft. to 4 ft. Peter Juen suggests that the higher the wall is to be, the shallower that the first lift should be, because concrete falling from a greater height exerts more pressure. If you walk around the house, by the time you return to the starting point the first lift will have had enough time to set a bit and won't transmit as much hydraulic pressure.

Consolidating the concrete well is critical with ICFs, so one way or another, most contractors vibrate the concrete. Some use so-

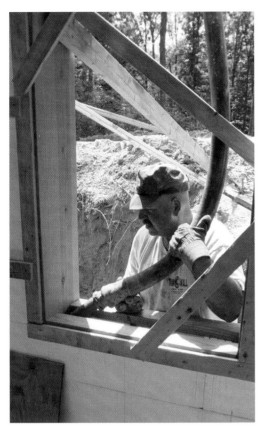

Pour the windows first. Pouring up to the bottom of the bucks first and moving on lets the concrete set. Pressure from adjacent concrete then won't cause the first pour to well up in the buck.

Sources

Most ICF manufacturers belong to the Insulating Concrete Form Association. Contact them for a membership list. For books and videos on ICFs, contact the Portland Cement Association. In researching this article, I found the website www.ICFWeb.com to be a great resource. It has a good discussion board and links to manufacturers of ICFs and their accessories.

Arxx Building Products
800 Division St.
Cobourg, ON K9A 5V2
Canada
800-293-3210
www.arxxbuild.com

IntegraSpec®
2743 Bunning Rd.
Sarsfield, ON K0A 3E0
Canada
613-835-2600
www.integraspec.com

Flexible Products Company®
1881 W. Oak Pkwy.
Marietta, GA 30062
770-428-2684
www.flexibleproducts.com

Insulating Concrete Form Association
1730 Dewes St.
Ste 2
Glenview, IL 60025
888-864-4232
www.forms.org

Portland Cement Association
5420 Old Orchard Rd.
Skokie, IL 60077
847-966-6200
www.portcement.org

ReechCraft Inc.
2001 1st Ave. N.
Fargo, ND 58102
888-600-6160
www.reechcraft.com

Todol Products Inc.
25 Washington Ave.
P.O. Box 398
Natick, MA 01760
508-651-3818
www.todol.com

called pencil vibrators, mechanical vibrators with thin 1-in. or ¾-in. shafts that don't hang up on webs and rebar. These devices take experience because mechanical vibration can blow out ICFs.

More commonly, builders use impact vibration, which means that they place a 2x4 over a web and then whack it with a hammer. Some hold a wood block over a web and vibrate it with the foot of a disbladed Sawzall® or with a palm sander.

Voids in the concrete are most likely going to be around windows and doors. The rebar and the fasteners that protrude from the bucks can create dams that keep concrete from filling all the spaces.

Most ICF contractors begin pouring at the windows (see the photo above). Because the buck bottoms are open, pouring higher than this level at once can cause concrete to well up through the opening. Pouring up to the bottom of the windows and then moving on gives this concrete time to set. It doesn't take long to stiffen enough that concrete placed nearby won't force the initial concrete out of the buck.

Next come the corners, both sides of which should be filled evenly to balance the pressure. On successive passes, it's important to fill evenly on both sides of bucks as well. Then, it's around and around the foundation until the forms are all full, the anchor bolts installed, and you're sitting in the shade with a cold beer.

*Except for a stint at Rutgers University, **Andy Engel** was a woodworker and carpenter from 1979 until joining the* Fine Homebuilding *staff in 1996. In 2004, Andy took a break from editing and spent most of the year working as a carpenter. He's now a senior editor at* Fine Woodworking.

A Fast Foundation for an Addition

■ BY RICK ARNOLD

Four years ago, a builder asked me to form and pour a crawl-space foundation for a small addition using insulating concrete forms (ICFs). I had never worked with ICFs before. He was a good customer, though, so I rolled my eyes and agreed. I grabbed my friend and fellow contributing editor Mike Guertin, figuring we could muddle our way through just about anything. Less than two hours later, we were ready to pour, and I was sold.

Since then, I have used ICFs to form everything from full basements to two-story houses. Along the way, I've discovered many benefits of building with ICFs (see the sidebar on p. 140), but I especially like them for small-addition foundations, like the 18-ft. by 18-ft. foundation featured here, because they can be installed so quickly. In fact, I've talked contractors and homeowners through the process over the phone.

Because ICFs are so easy to install, I can save a lot of money in labor by doing the job myself rather than paying a subcontractor. For instance, this job cost a bit more

than $1,400: $900 for the blocks and about $550 for the 7½ yd. of concrete. ICFs are available in several styles, each suited for different types of walls. I prefer the flat-wall type—even though it requires more concrete than other styles—because it emulates the more familiar poured-concrete wall.

My ICF of choice is called iForm® (see the photo on the facing page). The best thing about this ICF system is that because there is no top or bottom, the blocks can be installed in any direction; even corner blocks can be flipped to fit a right or left corner.

To estimate the number of blocks needed for a job, I start by dividing the length of each wall by the length of a single block. When added together, this tells me how many blocks I need in each course. Next, I divide the required height of the wall by the height of one block; this tells me how many courses I will need. When multiplied together, these two numbers give me the total blocks needed. I also like to add one extra block for each side of the foundation in case I run into problems.

Interlocking
teeth

Snap-in rebar
chairs

Expanded
polystyrene (EPS)

Embedded
fastening strip

Fastening-strip
indicator

Pros

- Unlike conventional concrete foundations, ICFs require no special tools or training. Basic carpentry skills are about all that is needed.
- The expanded-polystyrene blocks are light, easy to cut and place, and permanent.
- There is no hassle of stripping forms after the pour.
- Fabrication around difficult areas like live sewer pipes and electrical conduit is very easy.
- When concrete is poured inside ICF blocks, a negligible amount of water evaporation occurs, which is crucial to proper curing. Most contractors strip conventional forms a day or two after the pour, which increases evaporation and interferes with the curing process.
- Pouring concrete in cold weather is not a problem because the curing concrete generates heat that's retained by the insulation.
- If a conditioned crawl space is part of the building-envelope strategy, the ICFs provide a well-insulated and air-sealed barrier.
- Basements built with ICFs are less prone to moisture problems than poured-concrete walls.

Cons

- Exposed exteriors require a protective finish. I use an EIFS (exterior insulation and finish system) coating called Dryvit® or ¼-in. cementitious panels.
- The outer 2½ in. of the finished foundation is not load bearing, so I use a larger sill plate (2×8 instead of 2×6); I always double the plate anyway.
- Building officials may require the inside wall of an ICF crawl space to be covered with a fire-retardant material.
- ICFs offer no means of visual inspection for insects. Some brands of ICFs are treated for insects, but termite shields and ground treatment are good ideas.

Establish a Dead-Level First Course

To start, I sweep the standard concrete footing clean and snap chalklines representing the outside perimeter of the foundation. Next, I cut 2½-in. light-gauge steel-stud track to length and fasten it along the chalklines with masonry nails (see the bottom left photo on the facing page). The metal track, which normally is sold as top and bottom plates for steel-stud wall systems, just happens to be the right width to hold the iForm blocks firmly in place.

Before I start installing the blocks, I check the footing for level. The key to a quick, accurate installation is a dead-level first course. If the footing is more than ¼ in. out of level, I have some adjusting to do. This can be done by trimming the bottom of the blocks or by lifting them with shims, and later filling in any gaps with expanding foam.

Because each block is 16 in. tall and the overall foundation height often isn't a multiple of 16, I usually have to trim one course to get the correct finished height. If this is the case, I prefer to cut the bottom of the first course so that the top course is a full-height block. The blocks can be cut quickly to a uniform height on the tablesaw, using the rip fence as a guide.

Corners Come First

Because the corners are the weakest points in the foundation, they are the first blocks to be installed on each course. This ensures that I won't have to compromise their strength by cutting them to fit the rest of the blocks. After I lay the corners, I fill in with straight blocks.

As I work my way out of each corner, I usually have to trim one block on each side of the foundation to get the right fit. The iForm blocks have vertical lines at 1-in. increments on the sides, so I can take my jigsaw and cut by eye, using the nearest inch

The First Course Is Most Important

Anchor the first course of ICF blocks by fastening steel track along the chalklines with 1-in. masonry nails spaced about 2 ft. to 3 ft. o. c. I set the corners of each course first to eliminate any need to cut, and thereby weaken, this structurally vital part of the foundation.

Zip ties secure the corners. Because the corners are the weakest points in the form assembly, it's important to secure corner blocks to adjacent straight blocks with plastic zip ties.

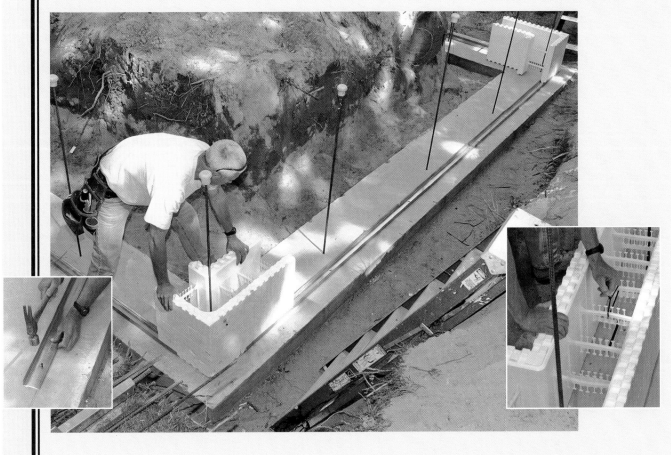

line as a guide. It's best to cut a little short rather than a little long; any extra length will push the corners out of square.

Block by block, I work my way toward the existing foundation. Butting the ICFs against a flat surface (poured concrete or concrete block) simply involves cutting the last blocks to the correct length and inserting rebar into the existing wall. But on this job, the existing foundation was fieldstone that had a very irregular surface. Thanks to the easy-working properties of the ICFs, I was

able to scribe the blocks to fit against the old foundation (see the top photo p. 143). I didn't need any rebar anchoring the new wall to the old because the rough fieldstone offers a strong bonding surface that poured-concrete and block walls do not.

Reinforce the Forms with Plastic and Steel

Once the first course is laid, I reinforce the corners by securing the corner blocks to

The key to a quick, accurate installation is a dead-level first course.

Rebar Strengthens the Concrete

Plastic zip ties and internal webbing hold vertical and horizontal rebar in place.

Rebar

Zip tie

Zip tie

Steel track

Concrete footing

After the first course has been laid, slice the side of one block and loop a zip tie around the vertical rebar that extends from the footing. Pull the tail end of the zip tie through the slot so that it's accessible from outside the form.

When I drop in another length of vertical rebar from above, it is carefully threaded into the looped zip tie, which then is pulled tight with a pair of pliers to tie the rebar together.

adjacent straight blocks with zip ties (see the bottom right photo on p. 141). Available at any hardware store, zip ties are thin plastic straps used for bundling cables and wires. I also use zip ties to tie together the two blocks that form the vertical seam. Then I snap in a row of ½-in. rebar and overlap it by at least 18 in. in the corners. The rebar chairs of the block I use are formed to allow two pieces of rebar (up to ⅝ in. dia. each) to be snapped together, one on top of the other, eliminating the need for wire ties. That's a real time-saver.

On the next course, I flip the corner blocks end for end so that the seams will stagger. I build up the corners pyramid style, making sure to attach the corner blocks to adjacent straight blocks with zip ties before stacking the next unit.

After measuring the diagonals of the foundation to make sure it's still square, I lay the rest of the straight blocks in place, reinforce the vertical seams, and scribe the ends as before. Staggering the cut seams in successive courses is not an option, because ICFs with interlocking teeth almost never line up correctly; therefore, one of the two rows of teeth would have to be cut off to get the forms to seat properly.

Handling Vents and Openings

To create a hole for crawl-space access, I build a permanent frame using pressure-treated lumber. The bottom of the frame is separated into two halves with a gap in the middle so that I can watch for voids underneath during the pour.

When I use steel vents meant for poured-concrete walls, I run into two problems. The first is that traditional vents are made for 12-in.-wide walls, but the ICF walls I use are 13 in. wide. This means the vent has to be recessed an inch on the inside of the crawl space. Also, there is no easy way to secure vents during the pour. I've managed to solve this problem by placing a heavy rock on top until the pour is finished.

Getting Ready for the Pour

I straighten the top of the walls with steel track, which keeps the interlocking teeth from clogging with concrete. I also string up a dry line about an inch or two above the outside perimeter of the walls. After snapping in the top row of rebar, I reinforce the vertical cut seams in the blocks by gluing them together with foam and screwing plywood scabs into the embedded fastening strips across both sides (see the bottom photo at right).

It doesn't take much concrete to move an unbraced wall, but it doesn't take much time to brace it either. As concrete is poured into the wall, it tends to push the blocks away from the existing foundation. To prevent this, I brace the ends of the wall opposite the existing foundation. Because the blocks I use lock together so tightly, bracing

Tie Up Loose Ends before the Pour

Even though the ICF foundation is strong and stable, I always reinforce large seams with a combination of plywood and foam, and I brace the outside corners to ensure a trouble-free pour.

Foam fills the gaps. Expanding foam can be used to seal seams, plug holes, and tighten rough scribe work.

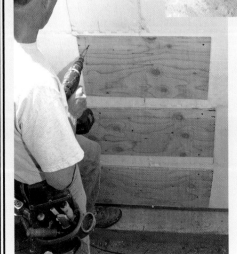

There's no such thing as too much bracing. It may be overkill, but I always brace the corners securely to prevent the blocks from pushing away from the existing foundation.

Plywood bandages. After gluing large vertical seams with foam, I reinforce them with plywood scabs screwed to the embedded fastening strips found on both sides of the seams.

Pour the Concrete

The pour requires close attention. It's best to pour the walls in 2-ft.- to 3-ft.-high sections called "lifts." Start with the corners before proceeding to the straight sections, making sure to watch carefully for a void-free pour. I use rocks to hold down the vents, and a dry line lets me double-check that the walls stay true.

Smooth the surface. Once the walls have been topped off, use steel track as a guide to remove excess concrete and to smooth out the surface before inserting anchor bolts.

the sides of the foundation isn't necessary. However, if you use another brand of ICFs, check the instructions.

The Pour Is the Trickiest Part

Most ICF manufacturers list how many blocks can be filled by 1 cu. yd. of concrete. For instance, 1 cu. yd. of concrete fills 7½ 13-in.-wide iForm blocks. So I simply count the number of blocks in the foundation, divide it by 7½, then round up to the nearest ½ yd.

The design and engineering of ICF blocks are based on a 3,000-psi concrete strength, but double-check your local code and the ICF manufacturer's specs to make sure you pour the right type of concrete in the foundation. I usually order 3,000-psi concrete with ¼-in. to ⅜-in. peastone aggregate. I've found that concrete with aggregate larger than ⅜ in. is difficult to pour and consolidate in the ICFs.

Pouring the concrete is the trickiest part of the process. Because the concrete needs to flow between the webbing of each ICF block, the consistency of the concrete is important. The lower the slump, the less water is mixed with the concrete, and the stiffer it will be. But a higher slump of 6 or 7 may result in substandard strength.

As with all three-wall foundations, I first anchor the corner blocks by filling them about halfway with concrete. Keeping the concrete at a 4 slump, I continue pouring the rest of the walls to the halfway mark. Then I have the truck driver loosen the concrete to about a 5 slump, start back at the corners again, and continue with the top half of the pour.

During the pour I constantly watch for voids forming, walls bowing, or gaps and seams opening up between the forms. I just use a stick to poke the concrete beneath the vents because I don't like to use a vibrator unless absolutely necessary. If gaps or seams open up, I stop the pour and reinforce the area with plywood scabs. Then I give the concrete a little time to set up, usually by continuing the pour in another area.

Once the wall has been topped off, I float it level with a trowel, using the steel track as a guide (see the bottom photo on the facing page). Then I install the anchor bolts. One last adjustment can be made simply by pushing or leaning on the wall to tweak it straight—and my foundation is formed, insulated, and poured in record time.

As a contributing editor for Fine Homebuilding, ***Rick Arnold*** *has written nearly a dozen articles related to framing. His most recent project involves installing four roof valleys on his new home.*

Sources

Founded in 1995, the Insulating Concrete Form Association (ICFA) is a growing group of professionals ranging from manufacturers to ready-mix suppliers, all focused on educating builders about the benefits of ICFs. Visit them online at www.forms.org.

The IRC (International Residential Code) has a section covering ICFs, complete with rebar tables. The source document for the code is "The Prescriptive Method for Insulating Concrete Forms in Residential Construc-

tion," published by the Department of Housing and Urban Development (www.hud.gov).

Dryvit Systems, Inc.
One Energy Way
P.O. Box 1014
West Warwick, RI
02893
800-556-7752
www.dryvit.com

Exterior insulation and finish system coating

Hardiplank
26300 La Alamed
Ste 250
Mission Viejo, CA
92691
800-348-1811
www.hardiplank.com

Cementitious panels

Portland Cement Association
500 New Jersey Ave
NW 7th Floor
Washington, DC 20001
202-408-9494
www.cement.org/homes

ICF manufacturers and distributors in your area

Reward® Wall Systems
800-468-6344
www.rewardwalls.com

Building a House of Insulated Concrete Block

■ BY FRED LEADBEATER

Stick-frame construction is all but dead in Europe, and for good reason. Wood is too scarce and too expensive to use as a hidden structural element, as it is in U.S. residential construction. Instead, European builders use wood primarily for decorative details. Masonry construction is standard, as I learned in years of job-related travel that took me to Europe. When I moved to Montana a few years ago and built a house, the technologies I had seen in Europe made more sense than standard wood-frame construction.

Weather conditions in the mountains of southwestern Montana are brutal: scouring winds, 90°F summers, and −40°F winters. The climate seemed ideal for one of the best European products I had seen, an insulated concrete block called Rastra®. Developed in Austria, the block is a type of insulated concrete form that is dry-stacked, reinforced with steel, and then filled with concrete. Eighty-five percent of the material's volume is recycled polystyrene, which is shredded into beads and mixed in forms with cement to make the block. Once completed, walls resist fire, earthquake, wind, water, and insects. Houses built this way are unbelievably comfortable in summer and winter, and pleasing to the eye to boot.

I am not a builder by trade, but my budget was tight so I joined the construction crew that consisted of the architect-builder Matthew Beardsley of the Baukunst Company in Bozeman and one other person. We used block from a now-defunct company called Ener-Grid to build my house. It seems identical to the block manufactured by Rastra. Another company is Cempo Forms Inc. Techniques we used to build my house should work with block from either of these companies.

Solid as a rock. Insulated-block walls reinforced with concrete keep the weather extremes of southwestern Montana at bay while offering protection from insect damage, fire, and earthquake.

Concrete for the floor and the first course of insulated block in the walls is poured simultaneously, creating a monolithic foundation for the rest of the building. Rebar runs from the footing into both the floor reinforcement grid and the walls for added strength.

Block is easy to cut and move around the site. Made of concrete mixed with recycled polystyrene, the block is easily cut with a chainsaw or a handsaw. A tractor and a pair of tongs provide the muscle to move block on the ground. But at about 160 lb., this 10-ft.-long block can be hoisted into place by hand.

Ener-Grid insulated block, 10 in. thick

½-in. rebar, 15 in. o. c.

½-in. PEX tubing for radiant heat, 9 in. o. c.

4-in. concrete slab

4-in. gravel

4-in. drainpipe set in gravel

Rigid-foam insulation, 1½ in. thick

Concrete footing, 8 in. by 18 in.

Rebar bent into floor reinforcement

Snap ties reinforce block. Standard snap ties run through 2×4 blocking will help to keep block courses aligned when concrete is poured. These second-floor walls need only modest bracing.

Blocks Are Lightweight, and Workable with Common Tools

Ener-Grid blocks were available in widths of 8 in., 10 in., and 12 in., and 15 in. or 30 in. high. Each block is 10 ft. long (sizes may vary, depending on which manufacturer you buy from). For all their bulk, however, blocks are relatively light. The 10-in. deep, 15-in. high blocks that we used weighed 160 lb. each. To move them, we used a rented farm tractor equipped with a simple clamp (see the bottom photo on p. 148).Once we were on the second floor, we also used a portable aluminum hoist.

One floor at a time. With the first-floor block filled with concrete, a ledger to carry the second-floor framing can be bolted into place. Bolts should be set before concrete is poured.

found that costs were comparable (my house cost roughly $65 per sq. ft. to build). Performance of the block, however, is superior to conventional construction in almost every way.

Block Walls Are Stacked, Foamed in Place, and Filled with Concrete

There's nothing unusual in the way these walls start: Our first step was to pour a concrete footing 18 in. wide and 8 in. deep (we used a frost-protected shallow foundation). Blocks are stacked directly on the footing. For a solid connection, we set lengths of #4 rebar into the concrete that would run up through the first few courses of block. When the blocks reached the level of the poured floor, Matthew used a chainsaw to cut notches on the inside face of the wall. Some of the rebar cast into the footing was bent 90 degrees and tied to the steel reinforcement for the floor (see the drawing on p. 148). When the floor was poured, walls and floor were connected in a monolithic structure. Radiant-floor heat is perfect with this technology.

Wall block is simply stacked up and temporarily glued in place with polyurethane foam. We used rebar both vertically and horizontally, running the steel into each hollow core after the walls were stacked. One of the beauties of working with the material is that holes can be easily patched with foam. So we could run horizontal rebar right through the side of the block into the core, then patch the hole with foam.

Outside corners are formed with 30-in.-high blocks mitered lengthwise at 45 degrees and set into place on end. That created a neat, straight joint. The blocks were easy to cut with a circular saw set to a 45-degree

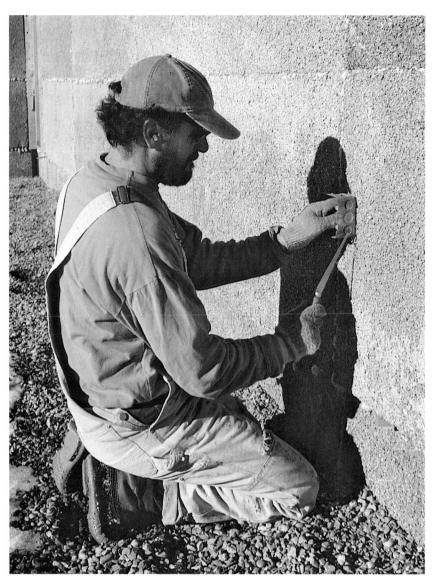

Wires and boxes are set into carved recesses. Because the block is relatively soft, many tools, including a keyhole saw, can be used to make electrical channels.

Houses built this way are unbelievably comfortable in summer and winter, and pleasing to the eye to boot.

The material is easily cut with a chainsaw or handsaw (see the top photo on p. 148). It can be routed, drilled, sanded, rasped, shaped, and set into curves. That's the real beauty of the material: Its applications are limited only by your imagination. Running through the block, both horizontally and vertically, are 6-in.-dia. cores 15 in. o. c. where steel-reinforcing bar and, ultimately, concrete are placed. Once filled, the blocks form a continuous, reinforced grid of incredible strength.

Before starting this project, Matthew had compared the cost of a well-insulated stick-frame house with Ener-Grid block and

A pump truck saves a world of trouble. A soupy mix of concrete is placed efficiently with a 5-in.-dia. line. Anchor bolts set in concrete at the top of the wall will be used to fasten a 2x top plate for conventional roof framing.

angle. Matthew cut one face, flipped the block to cut the other face, and finished the cut with a handsaw.

Windows, Doors, and Floors Require Special Planning

At window and door openings, we cut back the cores of the block to accommodate a 2x6 buck, or frame, whose faces were flush with the outer edge of the block (see the drawing below). These bucks prevent concrete from spilling out of the block as walls are poured, and they make a nailable foundation for installing doors and windows later. Bolts inserted through the 2×6 bucks and into the hollow cores of the block make

for a solid connection once the concrete has been poured.

We ran Ener-Grid block all the way up the gable-end walls, then snapped a line at the same angle as the roof pitch. Matthew cut to the line with a chainsaw, leaving core openings on an angle. We worried that the concrete would quickly drain out of these openings, but by making the concrete somewhat stiffer for those areas, we avoided a problem.

Floors can be hung at any height. Once the block walls were erected but before they were poured, we pushed anchor bolts through the block wall and into open cores. Later, we drilled corresponding holes in the wood I-joists that were used as ledgers for the second-floor framing and bolted them firmly to the walls (see the bottom photo on p. 149). We used spacer blocks to fill

Setting Doors and Windows in Place

Bucks, or frames, made from 2×6 material and anchored to walls before concrete is poured will provide a solid foundation for doors and windows. Bucks also prevent concrete from oozing out of open block cores.

Ener-Grid block

2×6 flush with inside of window opening

Anchor bolt, ¼ in. dia., countersunk in 2×6, 15 in. o. c.

½-in. rebar

2×6

Hollow core, 6 in. dia.

spaces between chords and webs. Although we could have used solid 2x ledgers here, Matthew had already determined the floor framing would be I-joists, and keeping joists and ledgers the same made it much easier to hang drywall ceilings down the road. Anchor bolts also were set into the top course of blocks so that a plate could be attached at the top of the wall for the roof framing.

The walls were reinforced with a combination of conventional snap ties and threaded rod bolted through 2x blocking (see the top photo on p. 149). We added bracing where we thought it would be most needed—at doors and windows, for example, and where the block was cut. All in all, bracing was on roughly 8-ft. centers and was used to straighten the walls prior to the pour as much as it was for strength.

A Pump Makes Placing Concrete Faster and Easier

With this kind of construction, the nail-biting comes when the concrete truck arrives. There are a lot of passages for concrete to fill, so we used grout with a slump of 8. That's roughly twice as fluid as standard ready-mix concrete. Matthew specified a seven-sack mix with 40 percent of the aggregate $\frac{3}{8}$ in. or less and the balance sand, which should yield concrete with a compressive strength of 4,200 lb. per sq. in. We also used a concrete pump truck, which simplified the pour (see the photo on p. 151). We placed the concrete with a 5-in.-dia. hose with an S-curve at the end. The curve is im-

You Will Find Little Agreement on R-Values for Insulated-Block Walls

By Scott Gibson

You might have better luck mediating Mideast peace talks than winning agreement on the thermal performance of insulated-block walls such as those in Fred Leadbeater's house.

Rastra, a manufacturer of insulated-block wall systems, estimates the dynamic, or effective, R-value of a 10-in. wall ranges from 19.7 to 23.9, depending on where you live (the dynamic R-value takes the effect of thermal mass into account). The steady-state R-value of the same wall is 11, says Rastra.

Castleblock®, distributor of Cempo Form block, is even more optimistic. It claims R-values of 3.5 per in., or R-35 for a 10-in. wall. The company says its numbers have been widely accepted by building officials. Besides, it adds, even if the numbers are high, 80 percent of a building's heat loss is through roof, doors, and windows, not exterior walls.

Energy Design Update (EDU), a monthly newsletter dealing with building-energy issues, popped those balloons in its December 1999 issue. The publication disclosed that tests at Oak Ridge National Laboratory found a static R-value of 7.6 for a 10-in. Rastra wall. That's less than a conventional 2×4 wall filled with fiberglass batts. When "dynamic benefit multipliers" are plugged in, EDU said, R-values rise to between 13.6 and 16.5, depending on climate.

Oak Ridge isn't ready to release a report. Although the lab has conducted a number of tests for Rastra, the work has not been finished.

If you have Internet access, check out Oak Ridge's Web page. It includes a wealth of information on the thermal performance of various building envelopes. At some point in the future, the lab hopes to add insulated concrete-block walls to the mix.

Sources

Cempo Forms, Inc
PO Box 9300
Pahrump, NV 89060
775-727-6565
www.cempo.com

Oak Ridge National Laboratory
PO Box 2008
Oak Ridge, TN 37831
865-574-4160
www.ornl.gov

Rasta
7621 E. Gray Rd.
Ste A1
Scottsdale, AZ 85260
877-935-3545
www.rasta.com

portant because it slows down the concrete as it leaves the hose. A gate on the S-curve allowed the person handling the hose to shut off the flow of concrete quickly.

Walls for the house were formed and poured in stages. We placed block for the first floor, poured concrete to that height and then added the second-floor deck. This process provided a solid platform for forming the remainder of the walls. Although Matthew has placed concrete in walls as tall as 17 ft. in one shot, a lower wall height reduces the risk of a blowout during the pour.

Any time wall height exceeds 8 ft., it's a good idea to place the concrete in two passes, or lifts. By the time you get back to the starting point, the concrete placed on the first pass will have stiffened somewhat. Because this concrete had such a high slump, we did not need to use any mechanical vibration to consolidate it. We used a total of 30 cu. yd. of concrete for my 3,000-sq.-ft. house.

With a Little Planning, Utilities Are a Breeze to Install

Plumbing and wiring are not as difficult as they seem. It is important to plan utilities from the start so that conduit and water supply lines can be located correctly in the concrete floor. From there, electrical and plumbing lines are run mostly through interior partitions. You do need receptacles on outside walls, and this is where you will be thankful that the block is easy to work with.

A channel can simply be gouged out of the wall to create enough room for standard plastic-coated cable. Just about any tool—router, keyhole saw, chainsaw tip—will do the job. Boxes are set in their own cavities (see the photo on p. 150) and then foamed into place. When the cable has been run,

the wire and channel are simply plastered over. We didn't think metal conduit was necessary.

When it came time to apply interior and exterior finishes, we rasped the rough block joints smooth with a thick piece of expanded metal lath attached to a handle. The material can be shaped easily. You can plaster or tile directly to the insulated block. Hanging pictures or other light objects from block walls is easy with a plastic anchor and screw. Heavier objects require more support. The best way is to plan by inserting bolts into the hollow cores before the concrete is placed.

Almost any siding can be used on the block. We used stucco on exterior walls up to the gable ends. From there, we hung cedar siding from furring strips routed into the block and screwed into the concrete core.

Fred Leadbeater is a former factory technician for Marker ski equipment and now lives in Bozeman, Montana.

Snug and comfortable. Dyed concrete floors that conceal a radiant-floor heating system and interior trim of vertical-grain Douglas fir help to give the house an earthy, low-tech appeal.

The Stay-Dry, No-Mold Finished Basement

■ BY ANDY ENGEL

Finished-basement projects usually begin with visions of a game room for the kids or of a secluded spot for Dad to watch Sunday football games with his cronies. Just about as frequently, these projects end badly, with black spots of mold, crumbling drywall, and a smell reminiscent of a dungeon. What goes wrong? In most cases, water becomes trapped behind a wood wall or floor and nurtures a bloom of rot.

However, it probably isn't a flood that causes the problem. Yes, bulk water, the kind that flows across the floor, needs to be eliminated before an attempt is made to finish a basement. (Some common measures include exterior waterproofing, functioning gutters, and/or an internal drain system.) But even if your basement looks dry, you easily can have problems when you enclose the concrete with a framed wall. The real

Continued on page 162

BEFORE

AFTER

A rec room that will last. Built with conventional wall framing, a plywood subfloor, and rigid-foam insulation, this basement remodel doesn't trap moisture that can cause problems later. (By the way, the stair railing isn't finished.)

2-in.-thick rigid foam
glued to rim joist

1×3 strapping,
16 in. o. c.

Plywood strip
blocks air.

Finish walls and ceilings
with ½-in. drywall.

2×4 stud wall

2-in.-thick rigid
foam glued to
foundation wall

Seam tape seals
joints between
panels.

Subfloor (two layers
of ½-in.-thick plywood
secured with concrete
screws)

1-in.-thick rigid
foam laid over
concrete floor

In summer, warm,
damp air infiltrates
and condenses on the
concrete foundation.

Vapor barriers can
cause damage by seal-
ing moisture in the
wall cavity, where it
promotes rot.

Expanded-polystyrene
rigid-foam insulation
is semipermeable and
won't trap moisture; it
also isolates the wood
framing from contact
with damp concrete.

Don't block moisture
with finishes such as
vinyl flooring or oil-
base paints. Instead,
use cork or wood floor-
ing, carpet, and latex-
base paints.

A waterproof founda-
tion that's adequately
drained prevents
flooded basements.

Understanding the Nature of Basements and Water

Moisture moves from wet to dry and from warm to cool. In the summer, damp soils and warm air outside make the moisture drive mostly inward. Humid outside air enters the basement and condenses on anything below its dew point: cold-water pipes, concrete walls, and floors. In particular, carpeted concrete floors can be a problem because they easily can become wet enough to support mold and dust mites.

Most basements dry out only in the winter, when interior heat sucks the available moisture out of the basement and drives some moisture outward through the exposed portion of the foundation. There's also some drive-out through the foundation itself because the basement is warmer than the surrounding soil. The trouble is that the soil tends to be wet, and so has a limited capacity for drying.

There's a significant energy cost in moving this water through the foundation. The traditional response has been to frame walls next to the foundation, fill them with fiberglass, and seal them with a plastic vapor barrier. But a basement vapor barrier can trap moisture and promote rot. Basement floors built with a similar system fare no better.

Plywood Subfloor Is Supported and Insulated with Rigid Foam

Use Rigid Foam Instead of Wood for Sleepers

To isolate the wood from moisture in the concrete, full sheets of EPS are laid on the concrete floor. Seam tape and expanding foam seal the seams against the infiltration of moist air.

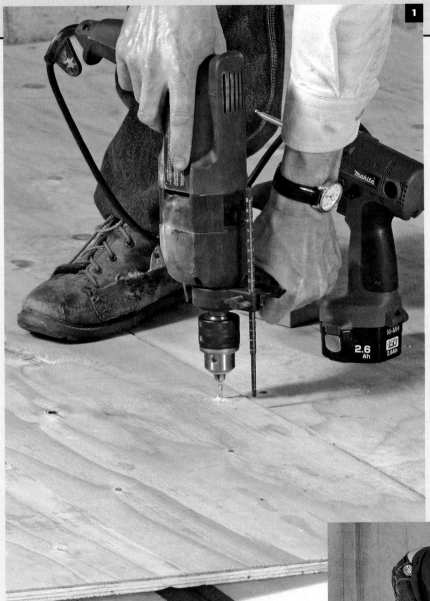

Two Layers of Plywood Go Down with Screws

After drilling and countersinking pilot holes (1, 2), the author attaches the first layer of ½-in.-thick plywood with 2½-in.-long concrete screws (3). To allow for expansion, ⅛-in. gaps are left between each sheet and around the room's perimeter. Laid at right angles to the first layer, a second layer of plywood is fastened with $1\frac{5}{8}$-in. drywall screws and spans the joints between sheets to make a stronger floor (4).

An Insulated Band Joist Stops Thermal Bridging

Insulating the joist bays with 2-in.-thick EPS will keep air leaks and cold spots to a minimum. Minimal amounts of expanding foam applied around the edges of each piece act as both a sealant and an adhesive.

Blocking air filtration

A strip of plywood is screwed to the underside of the floor joists. Extending from the edge of the mudsill to the inside of the wall plate, this plywood creates an air barrier. Gaps are filled with expanding foam.

Make the Foam on the Walls as Tight as Possible

After the floor is done, the walls are insulated with EPS sheets trimmed for a friction fit and glued to the foundation with expanding foam (1, 2). Seams and gaps are filled with the foam and taped. Plywood scraps keep the sheets in place until the glue sets (3). Unlike furring strips, a stud wall goes in plumb and straight, and allows room to run any utilities normally (4).

Sources

Building Science Corporation
70 Main St.
Westford, MA 01886
978-589-5100
www.buildingscience
.com

Information on Building Technology
The Energy Federation Inc.
40 Washington St.
Ste. 2000
Westborough, MA 01581
800-379-4121
www.efi.org

Foam gun, canisters of expanding-foam sealant, seam tape, and low-pressure dampers

Expanded-polystyrene (EPS) rigid foam is available at most lumberyards and home centers.

villain here is water vapor, the invisible moisture that keeps concrete damp and makes cold-water pipes drip with condensation in the summer. This water is always present. To reduce mold growth, water's contact with cellulose (paper, wood, etc.) has to be limited, and the water has to be allowed to escape.

Through research published by Building Science Corporation (see Sources), I've found that rigid-foam insulation both thermally protects the basement and breaks the contact between framing and concrete. To avoid trapping moisture, I never install a vapor barrier. Instead, I use materials and finishes that allow moisture to diffuse. You can get rid of this diffused water by installing a dehumidifier or by extending the air-conditioning ductwork into the basement. I'm no expert in this area, so let an HVAC contractor figure out the specifics.

Isolate and Insulate the Concrete

I use 2-in.-thick expanded-polystyrene foam (EPS, or Styrofoam®) on the walls and 1-in.-thick EPS below the plywood subfloor. This rigid-foam insulation is sufficient to make a noticeable temperature difference in the basement without crowding in the walls or the ceiling height. EPS is cheap, effective, and vapor permeable. Believe it or not, it also has the compressive strength to support a two-layer plywood subfloor without the use of sleepers.

After insulating the rim joist, I cover the floor with a layer of 1-in.-thick EPS. On top of this I lay the subfloor, then build a regular stud wall against the foam on the walls.

Keeping wood from contacting concrete is critical. Fail here, and you're inviting water in through capillary action. You could use pressure-treated plywood and framing lumber, but I think that's false security

and an unnecessary expense. If you've got enough moisture in the wall or the floor to cause rot, then you've also got the right conditions for mold growth, something that pressure-treated lumber won't prevent.

It's also possible to skip the stud wall and to screw furring strips to the concrete through the foam, but I don't like that approach for two reasons. First, I haven't seen many basement walls that are as plumb or as straight as I can build a stud wall. Unless you want to spend days playing with shims, the furring strips will mimic the defects of the foundation. Second, furring strips don't have the depth that allows easy installation of electrical boxes.

*Except for a stint at Rutgers University, **Andy Engel** was a woodworker and carpenter from 1979 until joining the Fine Homebuilding staff in 1996. In 2004, Andy took a break from editing and spent most of the year working as a carpenter. He's now a senior editor at Fine Woodworking. Any spare time is spent finishing his house, riding a bicycle, or leading Boy Scouts on rainy campouts.*

Air Quality and Fire Safety in the Basement

One of the effects of finishing a basement is to cut off leaks that may have been supplying combustion air for the boiler, furnace, or water heater. Failure to replace this air supply could contribute to backdrafting and the possible buildup of lethal amounts of carbon monoxide.

Unless the appliance manufacturer provides specs that say otherwise, the rule of thumb is to provide makeup-air ducts leading to the outside that are twice the size of the combined flues. In this basement, the boiler and water heater share a 6-in.-dia. flue. I provided two 6-in. supplies, one that ended at the ceiling level and one that ended near the floor in the mechanical room. To prevent these ducts from chimneying nice, warm air

to the outside, they were fitted with fabric dampers (see the top left photo below and Sources on the facing page) made for this low-pressure application.

One other safety consideration is basement egress. Most building codes require habitable basements to have two exits in case of fire. This basement already had two doors, so that requirement wasn't an issue. Lacking the second door, I'd have had to provide a code-approved egress window (a 5.2-sq.-ft. opening within 44 in. of the floor, leading to a 36-in. by 36-in. well with ladder rungs leading to grade). Last, if your home isn't already so equipped, install hardwired smoke and carbon-monoxide detectors in the basement.

Fabric damper

Combustion air

Outside air

Moisture-Proofing New Basements

■ BY BRUCE GREENLAW

Basements rarely conjure up cheery thoughts. They usually remind me of something out of the horror film *A Nightmare on Elm Street*. And when it comes to moisture problems, the truth about basements is a real nightmare. Surveys indicate that from 33 percent to 60 percent of the 31,000,000 basements in single-family homes in the United States have moisture problems.

A wet basement is uninhabitable, and the mold or mildew that thrive in such a musty environment can ruin anything stored there. Worse, a wet basement can wick more than 10 lb. to 20 lb. of water vapor a day into a home's interior. In today's airtight homes, this moisture can condense in the building envelope, causing mold, mildew, and eventually rot.

Basements don't need to be wet. There are scores of products that, when used appropriately and installed with care, help to keep basements dry. It's important to remember that these products are designed to function in addition to proper drainage. (For a detailed look at foundation drainage, see pp. 204–213.)

Waterproofing vs. Damp-Proofing

Some products are designed to waterproof foundations, while other products only damp-proof them. The American Society for Testing and Materials defines waterproofing as a treatment that prevents the passage of water under hydrostatic pressure. Damp-proofing, on the other hand, only resists the passage of water in the absence of hydrostatic pressure. Hydrostatic pressure is the force that is exerted on a foundation by the water that is in the ground that surrounds the foundation.

Determining the amount of hydrostatic pressure exerted on any given foundation is essential for choosing the appropriate foun-

Liquid-applied membranes are usually sprayed on. Liquid-applied polymer-modified asphalts such as Koch's Tuff-N-Dri® emulsion cure to form seamless, self-flashing waterproof membranes over concrete or masonry.

dation treatment. One way to predict the water-table fluctuations of a particular building site is to have a soils engineer perform an on-site soil test. A cheaper but probably less dependable way is to ask the neighbors.

Ultimately, the question is whether it's generally better to damp-proof or to water-proof a basement. *The Building Foundation Design Handbook,* published by Oak Ridge National Laboratory, recommends water-proofing for all habitable basements because most basements are exposed to at least some hydrostatic pressure. But if you're not planning to use your basement as living space, or if the soil around your house is exceptionally well drained, damp-proofing may be sufficient for the basement in your house.

Damp-Proofing Is Less Expensive but Not as Effective

Despite strong evidence that damp-proofing is insufficient in protecting basements from hydrostatic pressure, 95 percent of all builders damp-proof their basements, according to Koch Materials Company, which makes the Tuff-N-Dri basement water-proofing system. The success rate is questionable: 85 percent of builders questioned in a National Association of Home Builders survey said that at least some of the basements they have built leak, costing an average of $1,000 to $2,000 per callback.

The standard practice for damp-proofing new basements is to apply one or two coats of unmodified asphalt (asphalt with no chemical additives) to the exterior side of the foundation walls from the footings to slightly above grade. These asphalts come in various grades suitable for brushing, rolling, squeegeeing, spraying, or troweling. They cost roughly 40¢ to 60¢ installed per sq. ft. of basement-wall area.

Asphalt Emulsions Are Easiest to Work With

Water-based asphalt emulsions can be applied to damp substrates, including green concrete; they aren't flammable; they don't emit noxious fumes; and they clean up with water. They also can be used for gluing extruded polystyrene foam insulation (XEPS) to foundation walls. On the downside, emulsions must be protected from rain and freezing until they have dried, which can take several days in cool weather. Backfilling too soon can cause an uncured emulsified coating to deteriorate.

Cutback Asphalts Aren't Bothered by Weather

Cutback asphalts are solvent based and normally aren't affected by rain or freezing. But most cannot be applied to green concrete or to wet substrates. Uncured, they're toxic and combustible, and they dissolve foam insulation.

The trouble with unmodified asphalt in general is that it is a byproduct of the oil-refining process, and incremental improvements in refining technology gradually have eliminated asphalt's elasticity. Today's asphalts won't span the cracks that invariably occur in basement walls as concrete shrinks and foundations settle or move in response to various other conditions. Asphalts also tend to embrittle or emulsify with age, exacerbating the problem.

According to Brent Anderson, a consulting engineer in Fridley, Minnesota, if you plan to damp-proof with asphalt emulsions or cutback asphalts, it is best to apply it 60 mil to 100 mil, or just more than 1/16 in. thick. Gauges for measuring coating thickness are available from Paul N. Gardner Company Inc. Some builders improve the moisture protection of asphalt coatings by adding a layer of 6-mil polyethylene sheeting

Flexibility means better protection. Flexible cementitious coatings (left) might be called waterproof, but their limited elasticity can cause them to crack where foundations do. Elastomeric membranes (right), such as RPC's Rub-R-Wall® liquid-applied polymer, stretch to span cracks.

over the asphalt. Anderson says this treatment is an improvement but only if the top edges of the poly are carefully sealed to the foundation, and if the poly is protected from damage during construction.

Rubber Damp-Proofing Is More Elastic

Asphalt isn't the only material used for damp-proofing. One alternative is Rubber Polymer Corporation's new Rubber-Tite Damp Proofing Plus (for manufacturers' addresses, see p. 175), which is the only 100 percent rubber polymer damp-proofing I know of. This type of damp-proofing is much more flexible and better at spanning cracks than unmodified asphalt, but it degrades in sunlight and needs to be covered fast. It costs about 40¢ to 50¢ per sq. ft. installed, and it must be sprayed on by certified appliers.

Cementitious Coatings Are Durable but Don't Stretch

Polymer-modified cementitious coatings such as Masterseal® 550 (Master Builders Inc.) and Thoroseal® Foundation Coating (Thoro System Products) also are used for

damp-proofing. They cost about 20¢ to 65¢ per sq. ft. for materials, or about $1.25 to $2 per sq. ft. installed. These cementitious coatings occupy the gray area of moisture-proofing. Typically brushed or troweled on, they bond tenaciously to cured substrates and, in some cases, even stand up to hydrostatic pressure. They also breathe, which helps prevent basement condensation; they require no protection board; and they look good where they're exposed.

The trouble is, even the best polymer-modified cementitious coatings don't reliably bridge shrinkage and settling cracks (see the photo above left). Instead, they tend to crack where foundations do, admitting water where protection is needed most.

When cementitious coatings are used as part of a built-up system that includes rigid-foam insulation, mesh reinforcement, and synthetic stucco coatings, such as Sto's Below Grade System, they might truly waterproof foundations. But regardless of what manufacturers claim, cementitious coatings generally are considered by building codes as damp-proofing, not waterproofing.

Waterproofing Keeps All Moisture Out

If you elect to waterproof instead of damp-proof, there are several factors to consider when choosing an appropriate material. Products can differ in everything from shelf

Even the best polymer-modified cementitious coatings don't reliably bridge shrinkage and settling cracks.

life and compatibility with form-release agents to ease of application and cost. Some membranes require protection against backfill; others don't. Also, some membranes can be installed by the average contractor, while others have to be installed by factory-certified technicians.

Laminated-Asphalt Basement Waterproofing Is Similar to Built-Up Roofing

Laminated-asphalt waterproofing consists of two or more layers of a hot-applied or cold-applied unmodified asphalt, reinforced with alternating layers of a fiberglass or asphalt-saturated cotton fabric. Hot-applied laminated asphalts are seldom used today because they generate nasty fumes and pose a fire hazard during installation. Some companies, however, still promote cold-laminated asphalts as basement waterproofing. These asphalts are essentially the same ones used in damp-proofing, only they are installed in conjunction with fabrics designed to reinforce and augment moisture-proofing. Henry, for example, sells a fiberglass fabric for laminating its #107 asphalt emulsion.

The system has been approved by the city of Los Angeles for use as below-grade waterproofing, with the number of laminations used based on the amount of hydrostatic pressure.

Laminated asphalts cost less than $1 per sq. ft. for materials for a two-ply membrane. They span cracks better and are more durable than unlaminated asphalts, but they are inferior to many other waterproofing systems on both counts. Hot or cold, installation can be awkward, messy, and time consuming. Given the questionable long-term performance of unmodified asphalts, it might pay to take a hard look at the alternatives.

Liquid Applied Elastomeric Membranes Are the Most Widely Used Waterproofing Systems

Sprayed-on liquids that cure to form elastic membranes are probably the most popular basement-waterproofing products on the market, and for good reason. Not only can they be applied quickly to concrete or masonry (an experienced person can apply about 1,000 sq. ft. per hour), but they also cure to form seamless, self-flashing membranes. These membranes conform to complex surfaces, such as curved walls or walls that have a lot of lines or pipes passing through, and they span cracks up to $\frac{1}{16}$ in. wide. Also, because the entire membrane bonds to substrates, leaks are confined to small areas that can be detected and repaired easily.

Liquid-applied elastomers aren't perfect, though. Successful application requires painstaking preparation of substrates, meaning that voids must be filled with a non-shrinking grout or mastic; fins and lumps in the concrete must be removed; and surfaces must be clean and dry, or the membrane can blister or pinhole. For optimal performance, the thickness of the membrane must be monitored very carefully during application to ensure compliance with manufacturer's specifications (35 mil to 60 mil is typical). Also, some products must be heated and then sprayed on using expensive gear, which is bad news for do-it-yourselfers. In fact, many of these products require factory-certified application, although the products usually are backed by good warranties.

Polymer-Modified Asphalts Span Cracks Better and Last Longer

It's amazing what a little rubber can do for asphalt. The addition of rubber polymers dramatically increases asphalt's elasticity and longevity. At the same time, these sprayed-on modified asphalts are less expensive than many industrial-strength alternatives.

The two-part Tuff-N-Dri waterproofing system sold by Koch Materials consists of a polymer-modified asphalt emulsion, plus Warm-N-Dri® semirigid fiberglass panels. Installed by certified appliers only, the Tuff-N-Dri membrane is applied with an airless sprayer (see the photo on p. 164) at ambient temperatures down to 20°F. It adheres to green concrete and heals itself if punctured. The Warm-N-Dri panel protects the membrane from backfill, conveys groundwater to perimeter drains to prevent the buildup of hydrostatic pressure, and insulates, which on the exterior side of basement walls helps eliminate interior condensation. This system is ideal if you live in a cooler climate and plan to use your basement for living space. The system costs about 90¢ to $1.90 per sq. ft. (including labor), depending on your location and the system's R-value, and it carries a 10-year limited warranty.

Mar-Flex's® Mar-Kote Drain & Dry Waterproofing System is similar to Tuff-N-Dri, except that its modified asphalt contains aromatic solvents that evaporate completely and won't leach into the soil after backfilling. It can be applied at temperatures down to 0°F. This system costs about 90¢ to $1.20 per sq. ft., installation by certified contractors included. For the past seven years, Terra-Dome Corporation in Grain Valley, Missouri, has had excellent results using a hybrid system in the poured-concrete underground homes it builds. The heart of its system is a neoprene-modified asphalt emulsion made by Technical Coatings Co., called ADF-100. ADF-100 can be brushed, rolled, squeegeed, or sprayed. Once cured, it stretches up to 2,000 percent to bridge cracks. For extra moisture protection, Terra-Dome trowels ADF-500 mastic over cold joints before ADF-100 is applied, then lays strips of a bentonite sheeting called Paraseal® over the joint locations (more on bentonite later). The whole thing is covered with an insulating shell of XEPS foam before backfilling.

Paul Bierman-Lytle, a New Canaan, Connecticut, architect and builder whose specialty is nontoxic houses, swears by yet another modified asphalt called Safecoat® DynoSeal. Made by American Formulating and Manufacturing, it's a neoprene-modified asphalt emulsion that's designed for use by and for chemically sensitive people and for anyone else concerned with using nontoxic products. It can be applied only at ambient temperatures ranging from 45°F to 90°F, though, and costs more than the other modified asphalts: $1.12 to $1.20 per sq. ft., plus application. Bierman-Lytle has used DynoSeal in conjunction with a top-notch subsurface drainage system with great results.

Liquid-Applied Polymers Have the Best Stretch

Rubber Polymer Corporation's literature shows a guy trying to punch through a cured sample of the company's Rub-R-Wall waterproofing membrane. Instead of breaking, the green membrane stretches the length of his arm. Although the dynamics of a right jab and a cracking foundation are dramatically different, it is the stretchability of rubber polymers that is their greatest attribute.

Rub-R-Wall is heated and then sprayed by certified contractors over concrete, masonry, or rigid-foam foundation forms to a cured thickness of 40 mil. The membrane can be applied to frozen (but not icy) substrates at ambient temperatures down to 15°F. Drainage board or rigid-foam insulation can be

TIP

For optimal performance, monitor the thickness of the elastomeric membrane very carefully during application to ensure compliance with manufacturer's specifications.

attached to its sticky surface 15 minutes after application, and the surface stays tacky for several days. The bond is tenacious, though, so boards and panels must be positioned right the first time. (I know of one guy who used Rub-R-Wall to glue a detached heel back onto his boot.) Rub-R-Wall isn't UV-stabilized, so it needs to be covered promptly. Backfilling can proceed as quickly as 24 hours after application.

In its liquid state, Rub-R-Wall is flammable and toxic, but it's nontoxic when cured. Once cured, it stretches up to 1,800 percent to span foundation cracks (see the right photo on p. 167). Independent tests project a 100-year, below-grade life span for Rub-R-Wall, and RPC backs this up with a lifetime limited warranty. Installed price ranges from about 90¢ to $1.50 per sq. ft. RPC also makes a low-cost alternative to Rub-R-Wall

called Graywall. It stretches as much as 1,400 percent instead of 1,800 percent and is half the price of Rub-R-Wall.

For builders who can't backfill right away and where UV deterioration is a concern, Mar-Flex recently introduced a 100 percent polymer product, similar to Rub-R-Wall but with UV stabilizers, called Sunflex®. This product's bright-yellow membrane allows limited exposure to sunlight before backfilling. According to Mar-Flex, the installed cost of this membrane ranges from 70¢ to $1.15 per sq. ft.

Sheet-Applied Elastic Membranes Withstand Great Hydrostatic Pressure

Sheet-applied elastomers are best sellers on the commercial market. Homebuilders, however, generally use them for high-end work only. This usage is because these membranes withstand enormous hydrostatic pressure, which is typically not encountered in residential work, and consequently cost more than most of the other products I've mentioned. Prices can range from $1.50 to $2 per sq. ft. installed.

The most common sheet-applied elastomers are 60 mil thick and consist of a layer of self-adhering rubberized asphalt that's laminated to a waterproof polyethylene film on one side and covered by a protective release sheet on the other. Available in 3-ft.-to 4-ft.-wide rolls, the peel-and-stick membranes install vertically over primed substrates, adhering fully to localize leaks. Edges are sealed by special mastics.

Peel-and-stick membranes offer two layers of waterproofing to bridge cracks. The factory-controlled thickness helps eliminate the thin spots and holidays (skips) that result from spraying. Membranes don't pinhole, can be applied at temperatures as low as

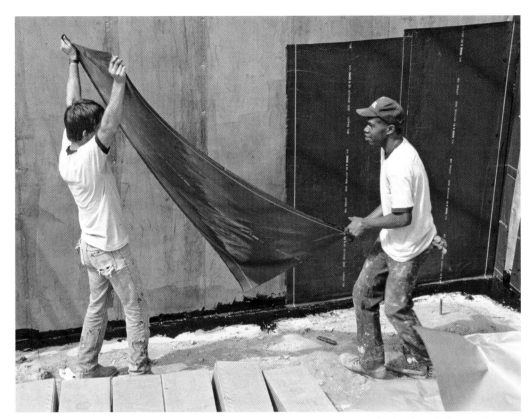

Peel-and-stick membranes stretch to bridge cracks. Self-adhering rubberized asphalt membranes such as W. R. Grace's Bituthene withstand tremendous hydrostatic pressure and can be backfilled immediately. Joints between footings and walls are covered with mastic or mortar for durability.

25°F depending on the formula, and can be backfilled immediately to eliminate jobsite delays.

On the flipside, peel-and-stick membranes don't conform well to complex surfaces, and they're difficult to flash. In addition, concrete substrates must cure for at least seven days before application and must be dry, thawed, and free of voids, fins, and other defects that could puncture the membrane. Also, time-consuming detailing is required at the inside corners and at the outside corners.

Once the sheets are applied, they can be difficult to reposition. They must be put on perfectly to avoid wrinkles, which need to be cut and repaired before backfilling. Protection board or matting also is required.

W. R. Grace's Bituthene® (see the photo above) has been on the market for about 25 years, so it's generally considered to be the standard peel-and-stick waterproofing membrane in the industry. But similar prod-

ucts are made by Karnak®, W. R. Meadows, Mirafi®, Pecora®, Polyguard®, and Polyken®.

The Noble Company sells a different brand of elastic waterproof sheeting. Called NobleSeal®, it's a chlorinated polyethylene (CPE) material that comes in 5-ft.-wide by 100-ft.-long rolls in thicknesses ranging from 20 mil to 40 mil. NobleSeal can be spot-bonded, applied to a grid of adhesive, fully adhered, mechanically fastened, or even loosely draped over foundation walls, spanning cracks up to ¼ in. wide (except when it's fully adhered). Seams are chemically welded. One distributor quotes the uninstalled price of NobleSeal at 60¢ to $1.20 per sq. ft., depending on thickness. Installation, however, can double that price.

Bentonite clay is mined from the Black Hills. Mameco's Paraseal is made up of granular bentonite clay, laminated to a tough, impermeable high-density polyethylene sheeting. Groundwater causes the bentonite to swell to six times its dry volume, which then forms a waterproof gel.

Bentonite Clay Is Nontoxic and Is Touted to Last Forever

Bentonite clay is another relatively expensive waterproofing material that handles great hydrostatic pressure and is used mostly for commercial or specialized residential work. The clay has been used in civil-engineering products since the 1920s and by builders since 1964. It's also used in toothpaste.

The key to bentonite's effectiveness is that it swells up to 15 times its dry volume when wet to form a sticky, impermeable gel in confined spaces. Not only is bentonite supposed to last forever, but it's also seamless, self-healing, and nontoxic. It also can be applied in cold weather. Most bentonite products can be installed with ordinary tools over masonry or green concrete with minimal surface preparation. Foundations can be backfilled immediately, in most cases without protection board, although a drainage layer might be required in some soils. Bentonite also can be an excellent choice for waterproofing slabs in extreme conditions.

On the downside, bentonite must be shielded from rain until backfilling, or it can wash away. Free-flowing groundwater can erode it even after backfilling. Also, some products have limited tolerance to soil salts, alkalis, and acids, though salt-resistant bentonite is available.

Bentonite comes in bulk for spray-on application or packed into 4-ft. by 4-ft. cardboard panels that dissolve after backfilling to leave a continuous bentonite membrane. Spray-on bentonite needs special equipment and skill to apply, and I've heard that the cardboard panels can leak at the seams before the cardboard disintegrates. I'd use bentonite sheets or mats instead (see the photo at left), such as the Mameco Paraseal that Terra-Dome uses. It comes in 4-ft.-wide rolls and features granular bentonite laminated to a tough waterproof high-density polyethylene (HDPE) film. Another alternative to Paraseal is Mirafi's Miraclay® matting, which costs about 60¢ to 80¢ per sq. ft. for the material or up to about $2 per sq. ft. installed.

San Francisco Bay–area builder Richard Kjelland, who has moisture-proofed foundations with everything from unmodified asphalt to Bituthene, tells me that bentonite sheets come in handy in tight spaces because they can simply be draped down foundation walls. This flexibility makes them a viable alternative where roomy foundation trenches are impractical. Paraseal can also be used extensively for waterproofing the outside of old rubble foundations, which are too rough and dirty to accept most other types of waterproofing membranes.

Dimple Sheeting Diverts Water while Moisture-Proofing

Relatively new to the United States but big in Europe, dimple sheeting is a low-cost waterproof membrane that doubles as a drainage mat. Delta®-MS sheeting, which is made in Germany and is sold by Intercontinental Construction & Equipment, is a 24-mil, dimpled HDPE sheeting that looks like an egg carton in profile. Available in rolls up to 8 ft. wide, it's simply rolled over concrete, masonry, or wood foundations and tacked up with special washered nails (see the photo at right). It can be installed over substrates in any condition and backfilled whenever you're ready.

The membrane not only repels water but also forms air gaps against the basement wall that will channel to footing drains any groundwater that might get through the membrane. These air spaces also allow the escape of indoor water vapor that condenses on the outside of the foundation. Depending on the quantity you order, Delta-MS costs 25¢ to 40¢ per sq. ft., and installation is supposed to take about one man-hour per 500 sq. ft. The Norwegian-made System Platon®, sold by Big "O" Inc. in Canada, is almost identical to Delta-MS.

Additional Products Augment Moisture-Proofing

As I've said, some moisture-proofing membranes need to be protected from backfill. Fiberboard and even roofing felt are traditional choices, but a wealth of innovative substitutes now give membranes more than just protection (see the left photo on p. 174), offering insulation and providing drainage.

For backfill protection only, Amoco®, Dow® Chemical, and U. C. Industries sell fanfolded XEPS panels that open into convenient, 4-ft.-wide by 50-ft.-long blankets. This foam blanket sticks beautifully to tacky, liquid-applied waterproof membranes such as Rub-R-Wall, or it can be glued with compatible adhesives to other properly cured membranes. Fanfolded protection boards are a measly ¼ in. or ⅜ in. thick, however, and therefore they can offer only a limited amount of insulation.

If it's an insulating-protection board you're after, all three companies also sell XEPS below-grade insulation panels (see the right photo on p. 174). Amoco's panels are available as much as 3 in. thick, while Dow's and U. C. Industries' come as much as 4 in.

Dimple sheeting provides drainage. This polyethylene sheeting resists moisture while the dimples provide air gaps that channel away any groundwater that might get through.

thick. These panels need their own protection where they extend above grade to prevent UV and baseball degradation. Retro Technologies makes a protective, stuccolike coating that can be brushed, rolled, or sprayed over the exposed foam.

Besides Koch's Warm-N-Dri and Mar-Flex's Drain & Dry, other products are available that protect, insulate, and provide drainage. Dow Chemical's Thermadry® is a 2-ft. by 8-ft. tongue-and-groove XEPS panel with a grid of channels on one side that direct groundwater to perimeter drains without affecting the panel's R-value. The channels are covered by a spin-bonded filter fabric

Dimple sheeting is a low-cost waterproof membrane that doubles as a drainage mat.

Thwarting condensation. An insulating board provides protection from backfill and helps eliminate interior condensation.

Protection, insulation, and drainage material work alongside waterproofing membranes. These boards and mats protect basement moistureproofing from potentially damaging backfill. The two mats on the right also direct groundwater to perimeter drains, while the middle three boards do that plus provide insulation. Left to right: Amoco Amocor-PB4 fanfolded, extruded polystyrene protection board; Koch Warm-N-Dri semirigid fiberglass insulating drainage panel (sold as a part of the Tuff-N-Dri system only); Dow Thermadry extruded polystyrene insulating drainage panel; GeoTech expanded polystyrene insulating drainage panel; Miradrain 2000R prefabricated drainage composite; and Akzo Enkadrain B drainage mat.

that admits water but keeps soil out. Panel thickness ranges from 1 in. to 2¼ in., with R-values ranging from R-4.4 to R-10.6. U. C. Industries' similar Foamular Insul-Drain® board comes in 4-ft.-wide panels.

The Insulated Drainage Board/Panel from GeoTech®, on the other hand, consists of expanded polystyrene beads bonded into a 4-ft. by 4-ft. panel that is available in thicknesses up to 2 ft. thick, although 2 in. is usually adequate for most residential applications. Groundwater drains between the beads. The panels are available with or without a filter-fabric skin.

Geocomposites are other products that protect foundations from backfill and hydrostatic pressure, but they don't insulate them. Most geocomposites don't look very sophisticated, but they've earned their stripes in tough commercial applications. The most common geocomposites consist of an impermeable dimpled plastic sheet with a filter fabric glued onto the dimples. The sheets measure ¼ in. to ¾ in. thick, come in 2-ft.-to 4-ft.-wide rolls, and are either nailed,

glued, or taped up with the dimples and filter fabric facing away from the substrate. After backfilling, the dimples and fabric team up to form silt-free subsurface waterways. Compressive strengths range from 10,000 lb. to 15,000 lb. per sq. ft., strong enough to restrain backfill without collapsing. Manufacturers I'm familiar with include American Wick Drain, Linq® Industrial Fabrics, W. R. Grace, and Mirafi. Price varies, but Mirafi's residential Miradrain® 2000R geocomposite costs about 35¢ to 45¢ per sq. ft.

Akzo®'s Enkadrain® is a different type of geocomposite that's quickly gaining ground in the residential market. The residential version, called Enkadrain B, is a 0.4-in.-thick mat composed of a systematic tangle of HDPE filaments bonded on one side to a filter fabric. It comes in 39-in.-wide by 100-ft.-long rolls and, where I live, costs about 30¢ to 35¢ per sq. ft.

*Price estimates noted are from 1995.

Bruce Greenlaw is a contributing editor of Carpenter *magazine and a former* Fine Homebuilding *editor.*

Moisture-Proofing Manufacturers

Here's a list of the manufacturers mentioned in this article. For a more comprehensive list, see Aberdeen's annual *Concrete Source Book* available from The Aberdeen Group. It's a directory that not only covers moisture-proofing but also just about everything else concerning concrete.

For names of local waterproofers, check the Yellow Pages, ask manufacturers, or call the Sealant, Waterproofing & Restoration Institute. The SWRI will also field waterproofing questions. If it can't answer a question, it will try to locate someone who can.

The Aberdeen Group
426 S. Westgate
Addison, IL 60101
800-323-3550
www.worldofconcrete.com

**Akzo Nobel® Inc.
US Headquarters**
300 South Riverside
Plaza, Ste. 2200
Chicago, IL 60606
312-906-7500
www.akzonobelusa.com

American Formulating and Manufacturing
350 W. Ash St.,
Ste. 700
San Diego, CA 92101
619-239-0321
www.afmsafecoat.com

American Wick Drain Co.
1209 Airport Rd.
Monroe, NC 28110
800-242-9425
www.americanwick.com

**Amoco Foam Products is now:
Pactiv Building Products**
2100 River Edge Pkwy.
Atlanta, GA 30328
800-241-4402
www.green-guard.com

Armtec
149 Thames Rd. West
Exeter, ON N0M 1S3
Canada
519-235-0870
www.armtec.com

BASF Admixtures
23700 Chagrin Blvd.
Cleveland, OH 44122
800-628-9990
www.masterbuilders.com

Convalence Adhesives
25 Forge Pkwy.
Franklin, MA 02038
508-918-1600
www.convalenceadhesives.com

The Dow Chemical Corp.
2030 Dow Ctr
Midland, MI 48674
989-636-1000
www.dow.com

GeoTech Systems Corp.
9912 Georgetown Pk.
Ste. D2
Great Falls, VA 22066
703-759-0300
www.geosyscorp.com

W. R. Grace & Co.
62 Whittemore Ave.
Cambridge, MA 02140
617-876-1400
www.grace.com

Henry® Co.
909 N. Sepulveda
Blvd., Ste. 650
El Segundo, CA 90245
310-955-9200
www.henry.com

Intercontinental Construction & Equipment Inc. (ICE)
7666 Highway 65 NE
Fridley, MN 55432
763-784-8406
www.incoeq.com

Karnak Corp.
330 Central Ave.
Clark, NJ 07066
800-526-4236
www.karnakcorp.com

Koch Materials Co.
6402 E. Main St.
Reynoldsburg, OH 43068
800-379-2768
www.kochmaterials.com

Mar-Flex Waterproofing Products
6866 Chrisman Ln.
Middletown, OH 45042
800-498-1411
www.mar-flex.com

W. R. Meadows Inc.
PO Box 7550
York, PA 17405
717-792-2627
www.wrmeadows.com

**Mirafi
A Division of Nicolon Corp.**
365 S. Holland Dr.
Pendergrass, GA 30567
800-234-0484
www.tcnicolon.com

The Noble Co.
PO Box 350
Grand Haven, MI 49417
800-878-5788
www.noblecompany.com

Pecora Corp.
165 Wambold Rd.
Harleysville, PA 19438
800-523-6688
www.pecora.com

Polyguard Producs Inc.
PO Box 755
Ennis, TX 75120-0755
800-541-4994
www.e-polyguard.com

**Poly-Wall®
Protective Coatings Technology Inc.**
5385 Frietag Dr.,
Ste. 1
Menomonie, WI 54751
800-846-3020
www.poly-wall.com

Retro Technologies Inc.
3865 Hoepker Rd.
Madison, WI 53704
608-825-9339
www.aceinfo.com

Rub R Wall
1135 West Portage Trail Ext.
Akron, OH 44333
800-860-7721
www.rpcinfo.com

Sealant, Waterproofing & Restoration Institute
14 W. 3rd St.
Ste. 200
Kansas City, MO 64105
816-472-7974
www.swrionline.org

Sto Industries
PO Box 44609
Atlanta, GA 30336
800-221-2397
www.stocorp.com

Thoro Systems Waterproofing
1850 Delmar Dr., #A
Folcroft, PA 19032
610-583-5502

Thrace-LINQ, Inc.
2550 W. Fifth
North St.
Summerville, SC 29483-9699
843-873-5800
www.thracelinq.com

Tremco Mameco International Inc.
3735 Green Rd.
Beachwood, OH 44122
800-852-9068
www.rpminc.com

Sealing a Crawl Space

■ BY LARRY JANESKY

Dirt crawl spaces are never-ending sources of moisture. Even if the dirt's surface is dry, digging down a little bit reveals moist earth. Moisture ruins houses by providing a hospitable environment for the fungi, mold, and insects that destroy wood framing. Moisture in a crawl space affects not only the floor system directly above it but also the entire house. Warm air in a heated building rises. As it rises, replacement air is sucked from the lowest part of a house. This natural air movement, called the stack effect, is how chimneys work. Consequently, whatever is in the air at the lowest point eventually flows through the upper sections of living space (see the drawing on p. 178). If mold spores and radon are present in the crawl space, you can bet they're in the living space as well.

Separate the House from the Earth, and Keep Out the Weather

With little headroom, light, and habitability, the crawl space may not seem to be an important part of the house. In fact, it's very important. Moist crawl spaces may be a bigger problem than wet basements because

they can produce an unseen moisture stream through the building envelope (see the right drawing on p. 178).

To avoid moisture's negative effects, a crawl space should be fully sealed and isolated from the ground and the outside. Part of my technique involves placing a 20-mil, 7-ply sandwich of high- and low-density polyethylene with polyester-cord reinforcement on the dirt floor and up the walls. I have this pool-liner-like sheeting made specially for my system. It is easily strong enough to crawl on and to store materials on. Its bright white color makes the crawl space a light, relatively pleasant place to be (see the photo on the facing page). A vapor barrier such as Tu-Tuf (Sto-Cote Products Inc.) also could be used, but it offers less durability, UV resistance, and fire resistance. Some contractors solve the problem of moist crawl spaces by pumping in concrete. But for this method to work, the concrete needs a vapor barrier. Even with a vapor barrier, this alternative doesn't address water-vapor diffusion through the walls. If the vapor barrier is doing all the work, then why use expensive concrete?

Despite what the building code says, many colleagues and I believe that venting crawl spaces is a bad idea. Code requires

Dramatic transformation. Proper detailing changed this crawl space from a spelunker's nightmare (inset photo) into a bright, clean, dry storage space.

If It's in Your Crawl Space, It's in Your House

As warm air rises inside the house, replacement air enters from the lowest part, often the crawl space. Properly sealing the crawl space and channeling rainwater away from the house can provide mold-free and radon-free living spaces.

PROBLEMS

A. Roof runoff and improper grading allow water to collect, where it can enter the crawl space through cracks in the foundation walls.

B. Humid summer air enters the crawl space through foundation vents and condenses.

Crawl space dampness can breed mold in living spaces.

SOLUTIONS

A. Roof runoff is directed away from the house with gutters and proper grading.

B. Sealing foundation vents keeps humid air out of the crawl space.

Dry crawl space discourages mold growth in living spaces.

C. Radon rises through the soil, into the crawl space, and ultimately into the living space.

C. A sealed crawl space keeps moisture from the ground out and passively mitigates radon. In tough cases, a PVC pipe and inline fan exhaust radon outside.

1 sq. ft. of ventilation for every 150 sq. ft. of dirt floor. Using a vapor barrier over the dirt floor reduces the ventilation requirement to 1 ft. per 1,500 sq. ft. of floor space. The intent is to vent out the humidity that the exposed earth lets in.

But venting creates its own problems. In winter, there's an energy penalty: cold floors and higher heating costs. In summer, vents actually admit moisture in the form of warm, humid air. Warm air can hold more moisture than cool air. Warm air entering a cool crawl space can reach its dew point and give up its moisture as condensation. Relative humidity, dew point, and the stack effect combine to make crawl space vents more likely to compound a moisture problem than to alleviate it.

For these reasons, I close the vents outside and seal them from the inside with 2-in. thick foam insulation and polyurethane caulk. And yes, I sleep well at night because I am doing the right thing for my clients.

Wet Crawl Spaces Need Drainage

In addition to water vapor, many crawl spaces leak groundwater. Such cases require a drainage system appropriate to the details of the crawl space. Outside, I make sure downspouts are directed away from the house.

Inside, I grade to one corner and install a sump and a pump with a sealed lid. Groundwater that leaks in can make its way from the dirt floor into the sump and be pumped out (see the top photo at right).

If a lot of water is leaking in, I create a swale in the dirt at the perimeter to channel water into the sump more directly. As another option, I can trench in a perforated pipe around the perimeter pitched to the sump. To prevent the pipe from clogging with silt, I slip a filter-fabric sleeve over it (see the bottom photo at right). Except in cases where there's extreme flooding or where a concrete

Got Water? Pump It Out

If gutters and exterior grading won't stop water from coming in, a trench, perforated drainpipe, and a sump pump may be required to move it out. A water-removal strategy could be one of the following scenarios.

PLAN VIEWS

If the ground slopes to one corner, install a sump in the lowest corner.

If the ground slopes to one side, dig a trench to channel water to the sump.

If the ground slopes more than one way, trench in a perforated pipe around the perimeter to carry water to the sump.

TIP

Never seal the liner to the sill plate, because doing so gives water vapor a path to the wood and can be a route for termites and other insects to get to the house framing. Leaving the liner 2 in. down from the top of the wall allows a routine termite inspection.

Plastic liner is easier to handle in the driveway. After the plastic is cut to size and rolled up, it can be installed in sections in the crawl space (see the top left photo). The author has liner material custom-made, but lighter plastic such as Tu-Tuf also can work.

floor will be poured, I avoid using crushed stone on perimeter drains because it's heavy and it's hard to lug through small openings.

I do one more thing before installing the liner: clean the dirt. All sharp or large rocks are buried when I regrade, thrown around the sump liner, or removed along with any wood or other organic material.

Installing and Sealing the Liner

Once the crawl space is cleaned out and drainage issues are solved, I install the liner. The liner material comes in a 24-ft. by 50-ft. accordion-folded roll. It's much easier to handle the 105-lb. roll outside than

To keep water and insects away from floor framing, the liner is cut in place 2 in. below the top of the wall.

in the crawl space, so I roll it out on the driveway to cut it to size (see the top photo on the facing page). I then fold and roll up the liner like a carpet, black side out, and bring it into the crawl space.

I start with a piece of liner wide enough to cover the floor from the center row of piers to the perimeter and up to the top of the wall. Next, I cover the other three walls, making sure the liner is long enough to overlap the floor by about 1 ft. Then I cut the remaining floor piece. After I roll out the liner, I take off my boots for the rest of the job to keep the liner clean.

Now, I turn back to the walls. I cut the liner 2 in. from the top of the walls (see the top left photo above) and fasten it

Holes are drilled through the liner into the foundation, and nylon expansion fasteners are driven into the holes.

Damp Crawl Spaces Can Make Your Living Room Moldy

Almost weekly, we see news stories about mold spores. Molds and individual tolerances to mold species vary. Some mold has little effect other than an unpleasant odor, but some can be toxic (most notably *Stachybotrys chartarum*).

Mold needs organic material, moderate temperatures, and moisture to grow. Remove one factor, and mold can't grow. Of these factors, moisture is the easiest to control. Mold thrives in damp environments. Using a hygrometer, we routinely find humidity readings well above 50 percent in dirt crawl spaces. Some are as high as 80 percent. This is compared with readings around 27 percent in conditioned spaces. According to the Environmental Protection Agency (EPA), relative humidity should be kept below 60 percent (ideally between 30 percent and 50 percent) to control mold. Properly sealing a damp crawl space lowers relative humidity to within these levels.

For more information on mold prevention, contact the EPA Indoor Air Quality Information Clearinghouse.

---

with nylon expansion fasteners (Outwater Plastics) that press into ¼-in. holes that I drill every 3 ft. or 4 ft. along the top edge (see the bottom photo on p. 181). I install the fasteners 3 in. down from the top of the liner (see the top right photo on p. 181) so that I can pull down the liner to seal it to the wall with polyurethane caulk (see the photo below). Polyurethane caulk is the only caulk I use on my projects; it sticks to anything and lasts for ages. I use Bostik 916 or Vulkem 116.

I never seal the liner to the sill plate because doing so gives water vapor a path to the wood and can be a route for termites and other insects to get to the house framing. Leaving the liner 2 in. down from the top of the wall allows a routine termite inspection.

At wood posts, I lift the weight with a hydraulic jack, if possible, and slide the liner under. If I can lift the post only a bit, I slide a piece of aluminum flashing under the post and seal the liner to the edges of the flashing with caulk. At masonry piers or columns, I cut slits and wrap the liner up 6 in. or so, then caulk. At seams, I caulk under the overlap and use a 4-in.-wide peel-and-stick tape made from a matching material, which makes the seams disappear. A high-quality builders' tape such as Tyvek tape also would work.

The finished system looks fantastic. My customers are pleasantly shocked when they see the end product. The white color reflects light, and it's clean and mold-free. Indoor air quality is improved, and the crawl space can be used for storage.

Seal the liner to the foundation with polyurethane caulk.

Larry Janesky holds 24 patents on basement waterproofing products and is president of Basement Systems in Seymour, Connecticut.

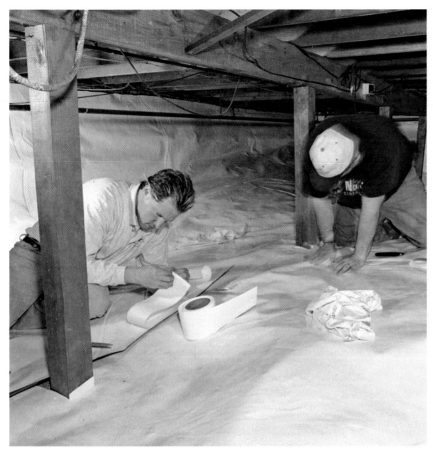

Seems like a single sheet, but it's not. Overlapping seams of the liner sections are sealed with caulk and matching tape for a seamless barrier.

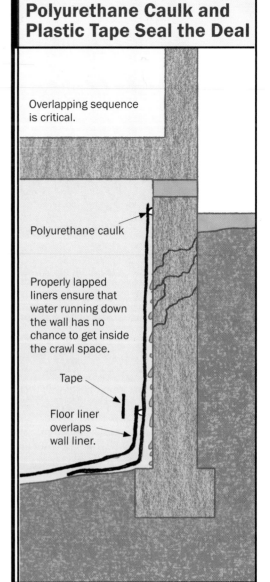

Polyurethane Caulk and Plastic Tape Seal the Deal

Overlapping sequence is critical.

Polyurethane caulk

Properly lapped liners ensure that water running down the wall has no chance to get inside the crawl space.

Tape

Floor liner overlaps wall liner.

Sources

Basement Systems®
60 Silvermine Rd.
Seymour, CT 06483
800-638-7048
www.basementsys-
tems.com

Bostik®, Inc.
11320 Watertown
Plank Rd.
Wauwatosa, WI 53226
414-774-2250
www.bostik.com

DuPont™ Tyvek®
800-448-9835
PO Box 80728
Wilmington, DE 19880
www2.dupont.com

**EPA Indoor Air
Quality Information
Clearinghouse**
P.O. Box 37133
Washington, DC
20013-7133
800-438-4318
www.epa.gov/iaq/
molds

**Outwater Plastics
Industries, Inc.**
24 River Rd.
Bogota, NJ 07603
800-631-8375
www.outwater.com

Tremco
800-321-7906
www.roancorp.com
tremcosealants.com
Vulkem® 116

**Sto-Cote Products,
Inc.**
P.O. Box 310
Richmond, IL 60071
(800) 435-2621
Tu-Tuf®

Details for a Dry Foundation

■ BY WILLIAM B. ROSE

As a research architect at the Building Research Council of the University of Illinois, I am paid to solve some of the more nagging problems that houses have. The most common problem I encounter is poor drainage away from the foundation. This problem became worse as wetlands were developed; I know what to expect when the name of the development is Frog Hollow.

I was once asked to look at a house that had settling problems. There was an addition, built over a crawl space, that was moving down relative to the main house. The dirt floor of the crawl space was low, even with the bottom of the footing. The soil along the edge of the footing was in small, rounded clumps, unlike the grainy, gritty surface of the rest of the floor. I dug away at the dirt clumps, and my fingers hit air. I dug a little more and found a space that reminded me of a prison escape tunnel. In all, 10 ft. of the footing was undermined.

Water from a downspout draining too near the corner of the house and the addition was the culprit. The water was taking the path of least resistance to the footing drain and sump pump in the basement of the main part of the house. That path happened to be under the addition's footing. By

following that path, the runoff had washed away the ground under the footing and caused the addition to settle.

Bad Drainage Can Cause a Raft of Problems

I call my studies of the zone where the house meets the ground "building periodontics." Proper preventative care of this area can avoid a variety of problems, some less obvious and a lot more serious than a damp cellar.

For example, a common problem in basements, particularly those with block walls, is inward buckling. This usually shows up as a horizontal crack one or two blocks below grade, or at windowsills, stepping up or down at the corners. A study I did with the Illinois State Geological Survey revealed the cause. Clay soils shrink during dry spells, forming a crevice between the soil and the foundation wall. Wind and light rains carry dirt into this crevice. Then, when seasonal rains come, the soil swells back to its original dimension, plus the increment of added soil. Over time, the wall is ratcheted inward

The Problems

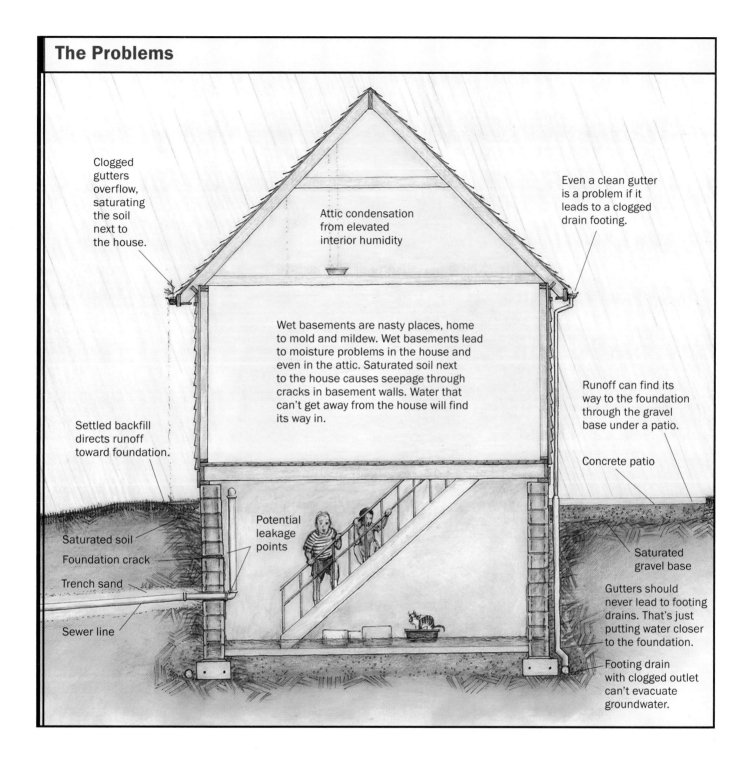

Clogged gutters overflow, saturating the soil next to the house.

Attic condensation from elevated interior humidity

Even a clean gutter is a problem if it leads to a clogged drain footing.

Wet basements are nasty places, home to mold and mildew. Wet basements lead to moisture problems in the house and even in the attic. Saturated soil next to the house causes seepage through cracks in basement walls. Water that can't get away from the house will find its way in.

Runoff can find its way to the foundation through the gravel base under a patio.

Settled backfill directs runoff toward foundation.

Potential leakage points

Concrete patio

Saturated soil

Foundation crack

Trench sand

Sewer line

Saturated gravel base

Gutters should never lead to footing drains. That's just putting water closer to the foundation.

Footing drain with clogged outlet can't evacuate groundwater.

and eventually buckles. You avoid this problem by keeping the soil next to the foundation dry.

Slabs suffer from water problems, too. Garage floors, for example, commonly crack at outside corners near where gutters drain. This cracking may be due to upward expansion of water directly below the corner. It can also be due to adhesion lifting of the perimeter wall, a situation occurring when saturated soil freezes fast to the foundation wall (see the drawing on p. 187). The soil nearest the surface is the first to freeze, and as the cold weather continues, deeper soil freezes. This saturated soil expands by 8 percent as it freezes, and it exerts a tremendous force that lifts the soil frozen to the wall above. The wall lifts and cracks the slab.

Clean gutter

French drain connected to the gutters at a distance from the house safely carries away runoff.

Clean-out

EPDM over 1-in. rigid foam

Plastic pipe

Details make the difference between a wet and a dry foundation. Every house doesn't need every detail shown here, but used together, they'd certainly keep a basement dry. Pay attention to soil types, the grade of the lot, and the climate, then pick the details that seem appropriate to the situation.

Swale intercepts water from uphill and diverts it away from the house.

EPDM over 1-in. rigid foam is a flashing system that keeps runoff from the foundation.

French drain

Concrete patio

Gravel

Tamped backfill

Trench sand

Impervious earth

Sewer line

Filter fabric

Gravel

Tamped surface pitched away from house helps water to drain.

Damp-proofing

Plastic pipe pitched to daylight or sump pump

Gravel

Filter fabric keeps the drains from clogging with silt.

Moisture damage around foundations isn't limited to masonry problems. In 1947, Ralph Britton, the government researcher whose work led to the current standards for attic-ventilation, showed that water vapor traveling upward from damp foundations caused most attic moisture problems. He concluded that if attics were isolated from wet foundations, the standard 1:300 venting ratio could be reduced to 1:3,000.

First, Pinpoint the Trouble Spots

Let's take a walk around an imaginary house and study the sources of its foundation water problems (see the drawing on p. 185). We easily spot the first one: The front gutters are clogged. Been clogged for so long, in fact, that saplings are sprouting in the composted leaves. Rainwater overflows these

Saturated Soil Leads to Frost Heaving

In a common scenario, water from downspouts has nowhere to go but next to the foundation. This results in damage to a garage slab from soil freezing.

Downspout

Cracked garage slab

Saturated soil, frozen fast to garage wall

Deeper, saturated soil exerts upward pressure as it freezes, lifting the wall and cracking the adjacent slab.

KEEPING WATER AWAY IS THE SOLUTION

Proper detailing can be as simple as sloping the ground away from the house and using downspout extensions to deposit runoff away from the backfill onto undisturbed soil.

Downspout

Garage slab

Extension

Splash block

Minimum 4 in. of gravel

1:10 slope

gutters, causing the ground below to settle. A small crater has developed, and its contents have nowhere to drain but down and into the cellar.

The gutters at the back of the house, however, are clean. The leader feeds into an underground drain that goes...where? Walking downhill, we find an outlet pipe at about the right elevation to be a footing drain. Composting leaves and granules from asphalt shingles clog the corrugated pipe. Might the water that should flow from this pipe be ending up in the basement?

A concrete patio, probably poured on a sand or gravel base, extends off the back of the house. A shallow depression next to the patio's edge collects a pool of runoff. This collected water will drain down the path of least resistance—through the gravel, under the patio, and down the foundation wall.

Going into the basement, we find leaks in places that confirm our observations. There is also a leak where the sewer line exits the house, indicating that water is flowing into the house through the sand in which the sewer line is laid.

Timing can provide clues to the source of leaks. If they occur in a matter of hours following a rain, the problem is surface water. If leaks follow a day or so after a rain, a rising water table is likely the cause.

Know Your Soil

As I write this article, I am sitting in the middle of the Midwest. The soil here is Drummer silty-clay loam, great for agriculture, murder on construction. The available water capacity is about 20 percent. This means that if I had 5 cu. in. of dry soil, adding 1 cu. in. of water would saturate it. Being clay, the soil will swell as I add water. The permeability is about 1 in./hr. This means that any layer of rain will need an hour to get through a horizontal layer of soil 1 in. thick. That's really slow. From these numbers, I can estimate that a 1-in. rain will saturate 5 in. of soil and take five hours for full penetration.

That's useful information. It is from the U.S. Department of Agriculture's county-soil survey, available from your county cooperative extension service agent. It allows a builder to estimate how much vertical water penetration there will be and how much of the rain runoff must be treated as sheet flow on the surface. This information matters a lot more on flat lots than on sloped ones, but it can still be important on the uphill side of sloping sites.

The perc test for septic systems is also a good predictor of how well the soil drains. If your soils have a good percolation rate, say, 10 in./hr. to 15 in./hr., to below the bottom of your footings, you may not have to do much to ensure a dry foundation. Install gutters and downspouts and make sure the first 10 ft. of ground around your house slopes away at something like 1 ft. in 10.

First Lines of Defense

What if your percolation rate is considerably less than 10 in./hr.? First, don't build on the lowest part of the lot, because that is where the water will go.

Gutters and downspouts are at the heart of rainwater management, the heart of moisture control in buildings. Deposit rainwater from gutters onto splash blocks and onto undisturbed soil so that the water runs away from the house.

Most modern houses are damp-proofed—that is, the exterior of the basement wall receives a bitumen coating. This provides a considerable amount of water protection. But water can enter through cracks resulting from utility penetrations, concrete curing, settlement, swelling soils, seismic activity, or other causes. Think of damp-proofing as a secondary defense against water.

Dealing with a Rising Water Table

Footing drains have been used for decades to intercept rising groundwater. Rising groundwater usually isn't the major problem for foundations. Surface water is much more likely to cause trouble if it isn't led away from the foundation. Still, footing drains should be installed. They don't cost much when you're excavating, and they're the devil to retrofit if you find later that you have a high water table.

Footing drains should consist of a foot or so of gravel around the outside of the foundation. Use a filter fabric over the gravel to keep it from clogging with silt. Filter fabric comes in several weights; the lightest is fine for residential use. Footing drains can have 4-in. perforated plastic pipe with the holes pointing down. They must lead to a sump pump or a gravity drain consisting of solid pipe leading to daylight. If you use pipe (as opposed to just gravel) in a footing drain, it should be slightly pitched toward the outlet, or at least not pitched backward. It should have surface clean-outs every 50 ft. Discharge by gravity flow is preferable to a sump pump. Sump pumps are a weak link, likely to fail when most needed, but a gravity drain may not be possible if the footing drains are deeper than any discharge point.

Gravel alone is probably just as good as gravel with pipes in it. A continuous gravel base that leads to a sump pump or to a daylight drain of solid 4-in. plastic pipe will handle most rising groundwater. The gravel is the main water route, so pipe used as a collector is not critical. Drain pipe here symbolizes good practice while making a doubtful contribution.

Never connect the downspouts to the footing drains, even if the drains run to daylight and not to a sump pump. Putting that volume of water closer to the footings makes no sense in light of my opening story.

Good Backfilling and Grading Are Crucial

Proper backfill procedures go a long way toward eliminating water problems. At the risk of sounding simplistic, be sure the ground

Planning for Your Building Site

There are so many soil classifications, foundation types, and climate variables that assembling general rules is challenging. If there is a general rule, it is this one: Design the soil surface that goes around the building to act as a roof. The overall aim is preventing the soil that is in contact with the building from being saturated with water. This "ground roof" should ensure that rainwater moves quickly and effectively away from the building. Downspout discharge, grading, flashing, drains, and surface soil treatment all play major roles in keeping the ground in contact with the building dry.

Basements, crawl spaces, and slabs all have their own peculiarities. With thoughtful design of the area where the house and the ground intersect, any foundation can be dry. Well, maybe any foundation that doesn't have provisions for boat docking.

slopes away from the house. You'd be amazed how many builders get this wrong. I recommend a slope of 12 in. in the first 10 ft. as a minimum (see the drawing on p. 190). Builders commonly don't allow enough extra soil for settling, and they almost never compact the backfill. Lightly compacted backfill may settle 5 percent of its height or more, often resulting in a situation where the grade pitches toward the house. When backfilling, include a correction for settlement. There aren't any hard-and-fast rules. Deep, lightly compacted backfill needs a big correction. Shallow, well-compacted fill might require none. Remember, too much slope near the house doesn't create water problems; too little does.

Moisture in the Ground Leads to Moisture in the House

Ground sloping to the house and no capillary break between the slab and the earth give water nowhere to go but up. Moisture from saturated soils will diffuse upward and warp floorboards, cause peeling paint and possibly rot the framing.

Warped floorboards

Moisture migrates up through concrete.

Downspout

Concrete slab

Saturated soil

Good detailing isn't much harder than bad. Ground that slopes away from the house and the gravel drains under the slab lead to a dry house.

Concrete slab

Flat floorboards

Downspout extension

Splash block

1:10 slope

Footing drain

Minimum 4 in. of gravel

6-mil. plastic

Crown and tamp the soil under the slab so that there are no ponds and so that it drains outward.

Compact the backfill as tightly as possible without damaging the foundation walls. Brace them well, using trusslike assemblies of heavy framing lumber spanning from wall to wall. Have the first-floor deck on, and fill all sides evenly. Block walls require more caution than poured concrete. Compact the backfill in 1-ft. lifts using a hand compactor, commonly called a jitterbug, or careful pressure from a backhoe. Because intersecting walls brace each other, the soil at outside corners can be compacted with less risk than in the middle of a long wall. For a minimum, compact these corners well, and be sure that all the downspouts drain near them.

Take special care where utility lines enter the house. They are frequently laid in sand that provides a direct path for water to reach the foundation. Be sure the soil under the utilities is well compacted, and cement and damp-proof the utility penetrations. Then use an impervious earth such as clay soil, or mix a bag of portland cement with the soil you have, to fill around the utility penetration. Tamp well (see the drawing on p. 186).

Slab Construction Needs Good Detailing Prior to the Pour

Getting water away from slab foundations is just as critical as with basements or crawl spaces. Remember, there are retrofit draining and venting options that can, to a degree, rescue a damp basement or crawl space. There are none that work on slabs.

Good preparation of the ground surface is critical prior to slab placement. Level the center, slope down to the excavation for the thickened edge of the slab, and compact the soil well. Pour the slab on 6-mil polyethylene over at least a 4-in. tamped gravel base. This base serves as a capillary break between the soil and the underside of the slab. Extend the gravel base to a footing drain to carry water away. It is important to remember that a capillary break works only as long as it remains unflooded.

Swales and French Drains

Other means of transporting water away from the house besides sheet flow (when the surface is effectively running water) are the swale and the French drain. A swale is a small valley formed by two sloped soil surfaces. Swales must be pitched, or they become ponds. A swale should be located away from the building, and it is often used to divert sheet flow coming from uphill (see the drawing on p. 186).

A French drain is a trench filled with rock or gravel that collects water and transports it laterally (see the drawing at right). I prepare the bottom of the trench so that it is smooth and carefully pitched toward the outlet. Mix dry cement with the soil in the bottom of the trench to make it less permeable, and fill the trench with whatever clean gravel is locally available. I hesitate to use road stone, a blend of gravel and stone dust, because water passes through it slowly. If the gravel is to be exposed, I try to cap it with

an attractive rounded stone. If the drain is to be covered, I provide graduated layers of smaller stone toward the surface, then perhaps filter fabric before the sod covering.

I sometimes use 4-in. smooth-wall perforated plastic pipe in a French drain, particularly if I expect it to carry a big volume of water, say runoff from the gutters. There are fittings that connect downspouts directly into this pipe. If you do this, install cleanouts at least every 50 ft. and keep the gutters clean. Otherwise, the pipe can become clogged with leaves. I don't use corrugated pipe for drainage because it is more difficult to ensure smooth, straight runs. It clogs more easily and is more difficult to clean out.

Concrete patios, stoops, driveways, and sidewalks abutting the foundation present problems. It is important to design them so that the gravel base beneath drains outward, a perfect use for a French drain. You may find that the driveway is one of the most convenient sites for a French drain. Driveways usually pitch away from the house, and a French drain can be integrated with the driveway so that it will not call attention to itself.

Utility lines that enter the house are frequently laid in sand that provides a direct path for water to reach the foundation.

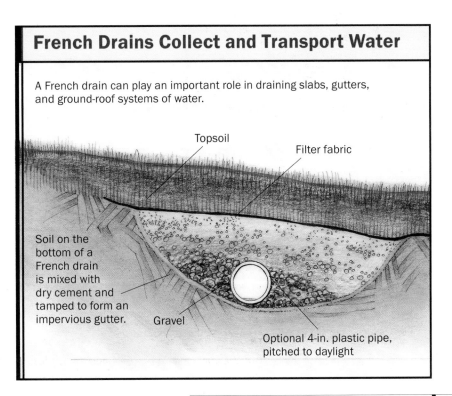

French Drains Collect and Transport Water

A French drain can play an important role in draining slabs, gutters, and ground-roof systems of water.

Topsoil

Filter fabric

Soil on the bottom of a French drain is mixed with dry cement and tamped to form an impervious gutter.

Gravel

Optional 4-in. plastic pipe, pitched to daylight

Where Should the Water Go?

To my knowledge, municipalities no longer provide storm-sewer service for new residential runoff. In my area, they do not receive the output from sump pumps. They receive and treat storm water to keep streets open, and that's about it.

If there isn't enough elevation difference between the house and a point on the lot where a pitched drain can come to daylight, then another solution is needed. Theoretically, if the pipe never pitches back, you don't need more than the diameter of the pipe in elevation difference. Practically, more is better, and ¼ in. per ft. is a good number to shoot for.

But say you don't have even that much pitch. For hundreds of years, cisterns and dry wells collected water below grade. Often, there was an overflow toward a leach field. Such design is still feasible, and indeed it is useful for garden irrigation where fresh water is scarce. Some municipalities require new subdivisions to handle runoff with on-site dry wells, rather than feeding it to a common detention basin. Usually a 1,000-gal. precast-concrete dry well (see the

drawing below) or commercially available plastic drainage structures are buried somewhere on site. Water from the gutters is piped in and stored until it can soak into the surrounding soil. The likelihood of success with either one of these systems depends on the perc rate of the soil and sufficient storage capacity to handle the maximum likely runoff. It also depends on how big the design rainfall is.

If you are not required to treat runoff in a specific manner, then take advantage of natural drainage courses on your lot. Get the water away from the house responsibly. If several downspouts connect to a French drain, enough water may flow from it to cause erosion problems. Place rocks under the end of the pipe and in the outwash area to spread the flow out and reduce erosion. Don't flood the neighbors' basements to spare your own.

Flashing at Ground and Foundation Intersections

I call that zone where the house meets the soil the "ground roof" because the soil surface must shed rainwater away from the foundation and the soil below. During dry spells, I commonly see a ½-in. crevice between the soil and the foundation. If that gap appeared on a roof, wouldn't we flash it? In severe cases or in old houses with hopelessly porous foundations, I have flashed this gap in the "ground roof" with EPDM roofing membrane (see the drawing on the facing page). Polyethylene sheets and bitumen membranes would work, but they degrade more easily when backfilled.

Ideally, I would flash a house as it was being built. In reality, I've done it only as a retrofit. A frost-protected shallow footing would lend itself well to a ground-flashing system. I dig down 8 in. at the foundation wall and extend outward 40 in., sloping the excavation 1 in. in 10. The hard work is the digging. Having done it, I should get

Dry Wells Avoid Draining Runoff on the Surface

Pipe from French drains or gutters

Filter fabric

Gravel

Required in some municipalities, dry wells give runoff time to soak into the ground. Their success is highly dependent on how well the surrounding soil drains.

Precast-concrete dry well

Flashing the House to the Ground Is an Innovative Means to a Dry Foundation

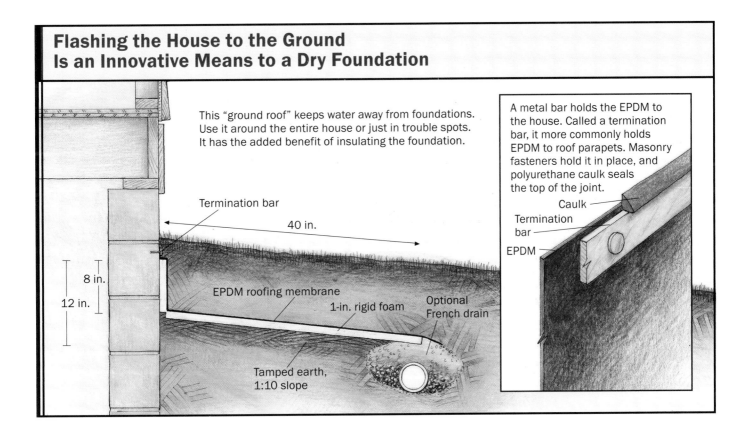

This "ground roof" keeps water away from foundations. Use it around the entire house or just in trouble spots. It has the added benefit of insulating the foundation.

A metal bar holds the EPDM to the house. Called a termination bar, it more commonly holds EPDM to roof parapets. Masonry fasteners hold it in place, and polyurethane caulk seals the top of the joint.

Termination bar

40 in.

8 in.

12 in.

EPDM roofing membrane

1-in. rigid foam

Optional French drain

Caulk

Termination bar

EPDM

Tamped earth, 1:10 slope

as much benefit as possible, so I take this opportunity to insulate the foundation. Slice through one side of a 4x8 sheet of 1-in. rigid foam, 8 in. from the edge, and fold along the cut. The resulting piece fits neatly into the 8-in. by 40-in. excavation. In the South, I suggest high-density mineral wool because it is less hospitable than foam to termites. In new construction, compact the backfill under the flashing well. Otherwise, settling could tear the EPDM from the wall or cause it to pitch toward the house.

After placing the insulation, I roll out the EPDM, letting it hang over the end of the foam. A metal strip called a termination bar, commonly used to attach EPDM to roof parapets, clamps the membrane to the foundation at grade level. I attach the termination bar to the foundation with expanding nail-in fasteners, alloy or plastic sleeves that slide into holes drilled into the foundation and then expand as a nail is driven in. I run a bead of cutoff mastic, a high-quality polyurethane caulk used for waterproofing termination bars on roofs, on this joint and backfill.

The flashing could extend farther outward from the building at the downspout locations. In existing buildings, you can often get away with flashing only the trouble spots, usually inside corners with downspouts. The "ground roof" need not be as watertight as a house roof. After all, moisture control in soils is always a matter of playing the percentages.

In soils with an average percolation rate, flashing by itself is enough to keep the water away from the foundation. If your perc rate is slow, install a French drain near the outboard edge of the flashing. Shallow plantings can go right on top of the EPDM.

William B. Rose is a research architect at the Building Research Council at the University of Illinois Urbana/Champaign.

In soils with an average percolation rate, flashing by itself is enough to keep the water away from the foundation.

Keeping a Basement Dry

■ BY LARRY JANESKY

I was standing in water a quarter-inch deep that covered an entire basement floor. The homeowner asked me in a surprisingly calm voice, "Is there any reason the basement of a brand-new million-dollar home should leak?" "Even if your house cost $100,000, it still shouldn't leak," I answered. I also told her that she had lots of company in her misery: A recent survey of 33,000 new-home owners revealed that 44 percent had leaky, wet basements. It was my guess that most of the basement problems were the result of a builder's neglect or efforts to cut costs.

As a basement-waterproofing contractor, I fix the mistakes of others. Having been a builder myself, I can empathize with the emotional struggle to "spend more and be safe" vs. "spend less, make a profit, we should be okay." However, no matter how much you spend, you shouldn't end up with a basement that seeps water like a cave; it's not good for the house or its occupants. Remember that it's much easier (and cheaper) to build it right the first time than to dig it up later with a jackhammer. To that end, I like to seal foundation walls, drain water away from the foundation's exterior, and expel the water that does manage to leak into a house's basement.

Sealing Foundation Walls

When water saturates the soil surrounding a foundation, it essentially creates a column of water whose cumulative weight increases as it rises in the backfill. This force pressing down is known as hydrostatic pressure; it drives water through joints, cracks, form ties, and other foundation imperfections. The first line of defense is some sort of exterior coating on the foundation walls.

These sealants are categorized into two groups: damp-proof and waterproof coatings. Damp-proof coatings are typically thin asphalt-based solutions that are sprayed or painted onto the foundation's exterior. The asphalt reduces the porosity of the concrete, but over time, it emulsifies in water and won't seal cracks. Some contractors mix fiberglass fibers with the asphalt to strengthen the mix but still offer only a one-year warranty. The low cost of damp-proofing (about 30¢ per sq. ft.) makes it attractive to many

Waterproofing, not damp-proofing, seals the foundation walls. More expensive than asphalt-based mixtures, a rubber-based membrane sprayed onto foundation walls remains flexible and waterproof.

builders, but the brief or nonexistent warranties (usually only a year) should make consumers wary.

Waterproof coatings, on the other hand, are a mixture of rubber and asphalt or all rubber (sometimes called elastomeric) and can cost three times as much. Like damp-proof coatings, waterproof coatings are sprayed onto the foundation (see the photo on p. 195), but the material must be heated before application; it's also applied in a thicker coat and is elastic enough to bridge ¹⁄₁₆-in.-wide cracks and small holes. The key to the waterproofing's performance is the amount of rubber in the mixture; more rubber means better performance and higher costs.

To protect any coating's integrity when the foundation is backfilled, many contractors cover the sealed concrete with what is known as protection board: sheets of fiberglass, rock wool, or extruded polystyrene foam. I use a ¼-in.-thick foam board that adheres to the fresh layer of waterproofing (see the photo above); at the very least, it provides a thermal break between the backfill and the foundation. Typical costs for a combination of waterproofing and rigid board can run from $1.10 to $1.50 per sq. ft.; 20-year warranties are common. Recently, a number of waterproofing manufacturers, such as Rub-R-Wall, came out with an extremely resilient coating that doesn't need protection board. But I still like to have the extra insulation provided by the foam.

When applying waterproofing, I make sure the joint between the footing and the wall is sealed. This means the top of the footing has to be clean before the wall is sprayed. Form ties should be knocked off both inside and out before spraying. It's also a good idea to determine the finish-grade height and spray to that line. If this elevation is miscalculated, 6 in. or more of untreated wall can end up beneath the backfill, causing leaks when the inevitable shrinkage cracks begin to appear in the foundation.

Rigid-foam protection board shields the membrane from backfill damage. Applied while the waterproofing is still tacky, the ¼-in. foam panels also provide a thermal break between the foundation and backfill.

Recently, a popular alternative to waterproofing has been waterproof drainage mats. The dimpled polyethylene sheets are unrolled and nailed onto the foundation wall, caulked at the top, and left open at the bottom. The mat's dimpled shape creates an airspace between the wall and the soil, so if water does leak in at the top or through a joint, it can drain to the bottom.

The problem with these drainage mats is that they must have an open footing drain below. If (or when) the footing drain clogs, hydrostatic pressure forces water up between the matting and the wall. Because the wall was never waterproofed, every crack and form-tie hole is vulnerable to easy water entry. In contrast, if a footing drain fails along a wall that was waterproofed, the form ties, wall cracks, and footing/wall joint are sealed and protected. However, even the best waterproofing guarantees only that water won't penetrate walls and doesn't prevent water from coming up around the footings and floor.

Keeping Footing Drains Clear

To keep a basement dry, you need to channel water away from the house with footing drains. Although most building codes say that foundations must have a drainage system of drainage tile, gravel, or perforated pipe, I always use 4-in.-dia. rigid-PVC pipe. (I've found that the black coiled, slotted pipe often used is difficult to keep straight; any dips in the pipe cause poor flow and clogs.) Two rows of ½-in. holes drilled at 4 and 8 o'clock positions keep the pipe's sediment intake to a minimum; slots will clog much faster than holes in most soil conditions. To make sure the pipe doesn't become clogged, I wrap the pipe with a succession of filters that resembles a giant burrito.

I start by cleaning the excavation to the bottom of the footing, usually with a shovel; a half-buried footing causes poor drainage. Next, I unroll 6-ft.-wide filter fabric along the footing, spreading excess on the ground away from the foundation and up the sidewalls of the excavation (see the drawing on p.198). I then dump 3 in. of ¾-in. stone on top of the fabric, level it off by hand, then set the PVC pipe so that at worst, it's level

around the entire foundation. Because the footings are mostly level, I'm happy if I can gain a few inches of pitch to the outlet.

During this stage of the project, it's convenient to tie the drains below each window well to the footing drains. Vertical risers made of solid 4-in.-dia. pipe run up from the footing drains under the window wells and terminate with a grate about 4 in. below the windowsill. The well can be filled later with stone so that leaves won't clog the grate.

I backfill over the footing pipe with more ¾-in. stone to an elevation of 8 in. above the top of the footing. As a rule of thumb, 1 yd. of aggregate will cover 12 lin. ft. of exterior footing drain; therefore, a house with 150 ft. of foundation will require a little more than 12½ yd. of aggregate. I pull the filter fabric up over the top of the stone and against the wall, using shovelfuls of sand or stones to hold it in place. If the fabric is not long enough to reach the wall, I add another course, overlapping by at least 12 in. Now the burrito is nearly complete.

Because the filter fabric will eventually clog, I put about 6 in. of coarse sand on top of the filter fabric. This progression of materials will keep the drain clear longer than any other practical way I know. The fabric protects the stone, the sand protects the fabric, and the soil won't wash into the sand.

As long as the lot's grade allows it, the exterior footing drains should always be run to daylight, pitched at ¼ in. per ft. or more, if possible. If the drains are servicing more than 200 lin. ft. of foundation, you might want to consider added measures. For example, you could put two outlets to daylight or increase the diameter of the outlet pipe from 4 in. to 6 in.

TIP

The black coiled, slotted pipe often used as a foundation's drainage system can be difficult to keep straight. Any dips in the pipe cause poor flow and clogs. Try 4-in.-dia. rigid-PVC pipe instead.

As long as the lot's grade allows it, the exterior footing drains should always be run to daylight, pitched at ¼ in. per ft. or more, if possible.

Clogged footing drains can't do the job. Water draining into the pipe carries silt that eventually fills the pipe. To avoid replacing a clogged footing drain, the author surrounds the pipe with a layered filter that stops sediment.

Making the Foundation's Exterior Impervious to Water

Gutter downspout

Because most water problems come from saturated backfill, it's important to seal foundation walls. An elastic rubber membrane sprayed onto the concrete seals small cracks, holes, and other imperfections. A PVC drain laid along the footing carries excess water away from the house. A multistage filter of stone, filter fabric, and coarse sand keeps the drain from clogging with silt.

4-in. PVC pipe buried approximately 2 ft. below grade

Window well

Grate is covered by stone in well to filter debris.

Vertical stack drains window well.

Foundation wall

Waterproof rubber membrane

A ¼-in.-thick rigid-foam board protects the membrane and creates a thermal break.

6 inches of coarse sand helps keep the fabric clear.

DRAINING DOWNSPOUTS AND WINDOW WELLS
If left unattended, rainwater that drains off the roof will end up in the basement. Gutter downspouts can be connected to 4-in. drain lines that run to daylight, away from the house. To keep window wells from leaking into the basement, a 4-in. vertical pipe run from a T in the footing drain can divert any water that collects there.

Outlet drain

A 4-in. rigid-PVC footing drain collects and diverts water to an outlet drain.

Footing

¾-in. stone

Filter fabric surrounds the stone and drain.

If there isn't considerable pitch on the lot or if a storm sewer isn't handy, the footing drains must run inside to a sump pump. A single 6-in. pipe that connects the exterior drain to the pump should be cast through the footing at the sump location and gives the drain as short a run as possible.

It's worth mentioning that it's never a good idea to connect an interior footing drain with a crossover pipe in the footing to an exterior footing drain that drains outside the house. The probability of failure on an exterior component of the system is high, while the probability of a failure to the interior system is low. If and when the exterior footing drain fails, the water will back up into the interior footing drain and flood the basement.

Drain the Gutters Far Away from the House

It may seem obvious, but many houses don't have adequate drainage for gutters and downspouts. Rather than carry water away with splash blocks, it's more efficient to connect the downspouts to a 4-in.-dia. PVC pipe (see the drawing on the facing page). Starting at about a 2-ft. depth at the house, the pipe should be pitched toward daylight as steeply as possible so that it can flush out the dead leaves and sticks that always accumulate.

Because gutters collect debris, it's a good idea to enlarge the downspouts to 3 in. by 4 in. instead of the usual 2 in. by 3 in.; this will also drain the gutter twice as fast in a heavy rain. The underground drain itself can be enlarged to a 6-in. dia. if necessary. In the drain, it's best to avoid 90-degree bends, which trap leaves and gunk, and use 45-degree bends instead. If there's a long run of gutter, give it two outlets. Finally, never connect the gutter drains to the footing drains, no matter how far downstream. Gutter drains are always voted "most likely to clog" and will back up the footing drains, too.

Getting Rid of Water in Basements

Although installing interior basement drains in new construction is a good idea, they're usually installed to fix problems in existing construction. Typical strategies include the use of interior perimeter drains that collect water from foundation walls, a vapor barrier below the slab that prevents water and vapor from wicking up through the concrete, and a good sump pump that will eject any water that collects in the drains and under the slab. In new construction, these methods all start with a new basement floor.

Under the slab, I like to lay 8 in. to 10 in. of clean ½-in. to ¾-in. stone. This stone allows the entire subslab area to drain and makes a good base for the slab. Once the gravel is laid,

Tough vapor barrier keeps basement slab dry. Made of 2-ply high-density polyethylene, the vapor barrier can be installed under or over the gravel base and keeps moisture from wicking up through the slab.

I always like to install a puncture-resistant vapor barrier (see the photo above). The strongest material is a 4-mil cross-laminated high-density polyethylene such as Tu-Tuff from Sto-Cote. It comes in 20-ft. by 100-ft. rolls that cost about 4¢ per sq. ft. I overlap adjoining pieces at least 18 in. and seal the lap with housewrap tape.

The concrete guys aren't crazy about the barrier because it doesn't allow the water in the mix to settle out, making the finishing process longer, but a 3-in. bed of sand laid over the barrier will alleviate the problem. (You can also lay down the barrier before the gravel is brought in.) Either way, the long-term benefits of an unbroken barrier under the floor are well worth the temporary inconvenience of installation.

Because water usually comes into the basement through the walls and the footing/wall joint, the best place to capture and

Never connect the gutter drains to the footing drains, no matter how far downstream. Gutter drains are always voted "most likely to clog" and will back up the footing drains, too.

Keeping the Interior Dry with a System of Drains and Pumps

To handle any leakage through foundation walls and to ensure a dry floor, an interior perimeter drain collects water from the walls and channels it to the sump pump. On sites where exterior drainage is poor, the pump can also be connected to the exterior footing drains.

Cross-laminated polyethylene vapor barrier prevents moisture from wicking up through the slab.

Pump drain discharges to storm drain or to daylight away from house.

Battery powers backup pump.

Perimeter drain diverts leaking water from the walls.

Concrete slab

Backup pump takes over during power failure.

Perforated-plastic sump liner

Sump pump

Stone bed beneath the slab drains into a perforated sump liner.

Interior perimeter drains are good insurance in new construction or retrofits. Plastic drains on top of the footing collect water that leaks in through the walls and channel it to the sump pump. The entire drain is to be covered with concrete, except the opening facing the wall.

channel the water is at the junction of the floor and wall (see the drawing on the facing page). The most efficient method is to install a plastic perimeter drain (photo above) that sits on top of the footing and below the slab; this perimeter drain collects any water that seeps in through the walls. In retrofit jobs, the first 6 in. of slab perimeter is cut away to expose the footing; once the drain is installed, the slab is patched. There are many manufacturers' variations on this theme; I use a drain that I designed and now

manufacture. Rectangular in cross section, it has a slotted opening facing the wall that channels water down and out toward the pump. The interior footing drain costs about $30 to $40 per ft. in a retrofit.

A less expensive version of this system for new construction is known generically as a draining floor edging (see the photos on pp. 202–203). It's a 6-in. by 4-in. L-shaped plastic flashing with a dimpled design. The dimples are laid against the wall and footing side and allow water to pass into the gravel

Drainage flashing is a simpler, less expensive interior drain. Installed at the junction of wall and footing, this flashing has a dimpled profile that allows any water from the walls to flow beneath the slab and into the sump pump. Obviously, it's important to keep the top of the flashing clear when pouring the slab (photo, facing page).

below the slab, where it can be pumped out. The flashing material costs about $1.50 per ft. You can find versions of these products through your local waterproofing contractor, or you can contact my company, Basement Systems®.

Sump Pumps for Interior Drainage

Any drainage technique I've mentioned is dependent on one thing: a good sump pump. After installing more than 10,000 pumps myself, I have some criteria for choosing a pump. First, it should have a cast-iron body; second, it should be able to pump ½-in.-dia. solids; and third, it should have a mechanical float switch (a float riding up and down on a rod), not a pressure switch or a "ball on a wire" design that

often hangs up. A Zoeller® M-53 fits the bill perfectly, costs about $150 and pumps 2,600 gal. per hr.

The pump sits inside a plastic bucket called a liner that collects water and separates it from the surrounding dirt so that the pump can push it out. The liner should be perforated and set in a bed of aggregate. Any drainage system that feeds water to the sump should have a pipe cut through the sidewall of the liner. The liner should have an airtight lid that will keep moisture from evaporating into the basement; it also keeps objects from falling into the sump hole that could interfere with the switch operation.

Relying on a lone sump pump to keep the basement dry is risky. It's all too common for a storm to knock out the power and flood the basement all in one night. A battery-operated backup pump provides insurance and often is equipped with an alarm that announces a pump failure. The backup can usually be installed in the same sump hole as the primary pump and use the same discharge line. The best backup systems use pumps that sit up off the sump floor, have float switches, and use matched chargers and batteries made specifically for long-term standby use. This last item is critical. Many backup units don't provide a battery; if the battery and charger aren't matched, the system won't charge properly. Backup-system equipment can cost from $300 to $1,000.

Discharge from the sump pump should be piped to a storm drain if one is available or to the exterior, where it will flow away. The discharge line should be installed so that it doesn't freeze during the winter.

*Price estimates noted are from 2001.

Larry Janesky holds 24 patents on basement waterproofing products and is president of Basement Systems in Seymour, Connecticut.

Sources

Basement Systems®
60 Silvermine Rd.
Seymour, CT 06483
800- 541-0487
www.basementsystems.
com

Rub-R-Wall of Ohio Inc.
6295 Cosgray Rd.
Dublin, OH 43016
614- 889-6616
www.rubrwall.com

Sto-Cote Products, Inc.
P.O. Box 310
Genoa City, WI 53128
815-675-2358
Tu-Tuff

Wall Guard Corp.
7024 N. Honeytown Rd.
Smithville, OH 44677
800-992-1053
www.wall-guard.com

Zoeller Pump Co.
3649 Cane Run Rd.
Louisville, KY 40211
800-928-7867
www.zoeller.com

Foundation Drainage

■ BY DAVID BENAROYA HELFANT

In a hillside community of homes I inspected recently, I found an interior stairway that rests on a "floating" concrete slab-on-grade. The owners of the house complained about the constant repairs needed to patch the drywall and baseboards that were connected to this stair. It seems that every time it rained, the stair would move up relative to the rest of the house. When the ground dried out, the stair sank.

The problem was that the stemwall foundation around the house was poorly drained, resulting in an inconsistent moisture content in the soil under the footings and allowing runoff to migrate beneath the slab some distance from the exterior stemwall. The slab was placed atop the undrained clay-laden soil, and there was negligible weight on the slab. Every time it rained, the clay expanded, taking the stairway for a short ride upward, tearing the drywall joints, and wracking the baseboards, railings, and casings out of alignment. The remedy to the situation was the placement of a drain uphill from the slab that diverted the water away from it.

The objective in designing and installing drainage systems around the perimeter of a foundation is to keep water from soaking into the soil and moving under the footings.

Water initiates the undermining of a foundation by causing erosion beneath it, literally carrying away the soil upon which the footing bears. Often this is the prelude to building settlement. If the soils have a high clay content, poorly drained foundations can be cracked or rotated by forces exerted by the wet clay as it expands (see the photo on the facing page).

If the grade around a house is well-sloped, you may not need a subsurface drainage system. But if your house is on a hillside made up of soils that drain poorly, such as clay, subsurface drains can be essential to the long-term well-being of the structure.

In most cases, a structure won't be threatened with a terminal illness brought on by bad drainage, but it can suffer an abundance of minor maladies. A damp crawl space can cause mustiness, mold, and mildew in a house, and fungus wood rot and termites thrive in this kind of environment. Soils under foundations that undergo dry/soggy cycles can bring on the familiar phenomenon of sticking doors and windows. While good site drainage may not solve all moisture problems (such as condensation), it can be effective in combatting cyclical changes, such as the floating-slab phenomenon described before.

Good drainage is essential for foundations built on heavy clay soils to avoid cracking and rotating.

System Basics

An effective drainage system consists of two distinct systems: a subsurface drain to carry away the flow of ground, or subsurface, water, and a surface drain to convey rain or snowmelt away from the building (see the drawing below). The core of any subsurface drainage system is a network of perforated pipes laid at the bottom of trenches next to or near the foundations, and sloped to drain toward a suitable receptacle. The pipe is laid with the perforations pointing down, so that water seeping into the trench from below will rise into the pipe and be carried off. Above the perforated pipe is a run of unperforated pipe (see the photo on facing page) that is used to transport runoff from roofs, patios, walkways, and other paved surfaces. Typically, these pipes will lead to a dumping site 10 ft. to 20 ft. downhill from the house (more on this later).

You may ask, "Why can't I just run my downspouts into the subsurface drain, and do away with the surface drain?" Don't do it. Combining the two increases the potential for a clogged line, and it defeats the purpose of a subdrain by injecting water into the ground.

HOW MUCH FLOW WILL DETERMINE THE TYPE OF PIPE TO INSTALL

For most residential drain lines—both surface and subsurface—we use 3-in.-dia. pipes. But if we're working on a hillside where we expect heavy flows, we'll install 4-in. subsurface lines. If a large roof area is draining into a single downspout, we'll play it safe and install a 4-in. surface drain line to carry the runoff.

Pipes and fittings for drain lines are quite similar to those used for DWV (drain, waste, and vent) work, but the fittings don't come in as many configurations and the pipes aren't as heavy. They also cost less—about 40 percent to 50 percent of what corresponding DWV materials cost.

We prefer to use smooth-wall pipe and fittings made of polyethylene plastic for subsurface and surface lines. This is a fairly rigid pipe that can be cleaned by an electric snake without being diced up from the inside out. We specify pipe that is rated at 2,000 lb. of crushing weight. This is important because

Section through Drain Lines

Patio sloped to drain

Area drain

Leader from gutter

Cleanout

Surface drain line

Sealed transmission line

Filter fabric

The depth of the trench varies according to the depth of the footing. To get positive drainage, the subsurface line may be a few inches below the footing, so long as its trench doesn't undermine the footing.

¾-in. rounded gravel

12 in. to 18 in.

Footing

Subsurface drain line

Two pipes. To properly drain a structure, you need to pick up subsurface water moving through the soil as well as runoff carried from flat surfaces and roofs. The pipe in the bottom of the trench above is perforated, with its holes oriented downward. Gravel covers it to within a foot of grade, and filter fabric is folded over the gravel to keep fines from clogging the drain line. The top pipe carries surface runoff from downspouts and drains. The pipe stub projecting above the tamped earth will attach to an area drain.

subsurface pipes are often buried well beneath the surface, and we usually compact the earth above them. Where the line passes under a sidewalk or a portion of a driveway that will carry traffic, we switch to cast-iron pipe and link the two materials with no-hub couplings.

When we can't get the polyethylene pipe, we use polystyrene pipe instead. But this material is brittle, which makes it tougher to assemble the fittings and pipe sections. We never use clay-tile pipe, which comes in 12-in. to 16-in. sections that butt against each other. There are too many opportunities for sections to move differentially. Nor do we use the thin, corrugated polyethylene pipe, because an electric snake can rip it apart if a line needs augering to clear a blockage. Its weak walls make it suspect for deep trenches, and its fittings do not seem to seal well.

In deep systems where it is necessary to carry large volumes of water, corrugated galvanized-steel or thick-walled ABS or PVC pipe may be preferable. Under these conditions, you should consult a geotechnical engineer for specific recommendations regarding dimensions and types of pipe.

INCLUDE CLEANOUTS FOR EASY CLEANING

All of our drainage systems are designed with cleanouts, similar to conventional wasteline plumbing systems, so that the system can be cleaned with an electric snake. We put them at 30-ft. to 40-ft. intervals on straight runs and at strategic locations elsewhere: major bends, intersections with other lines, and the point at which downspouts enter the surface-water drain system (a prime target for a leaf clog). We always cap the cleanout with a plug so that it does not collect debris.

We handle surface-drain flow either with an area drain, a catch basin, or a trench drain (see the drawing below). The first two are concrete, alloy, or plastic boxes that have metal grills to keep debris out of the systems. The area drain is connected to a leader that ties into the surface drain line (see the drawing on p. 206). A catch basin collects water from several surface drains and feeds it into a single outlet. A catch basin is also deep enough to allow sand and soil to fall to the bottom, where they collect without interrupting the water flow. These "fines" settle into a sludge that should be removed periodically. Trench drains are long, narrow steel, plastic, or fiberglass boxes with grills on them. You see them at the base of driveways, where they catch the water before it inundates a garage that's downhill from the street. We usually install the ones made by Polydrain® (ABT, Inc.®).

Tools of the drainage trade are neither mysterious nor high-tech. They include picks and shovels as well as pneumatic and electrical demolition and digging tools to break up the earth. Good wheelbarrows are essential. We recently acquired a Takeuchi® tractor, which has a backhoe and an auger attachment. It speeds up excavation consid-

Surface Drains

Area drain

Trench drain

Bottom sloped to drain.

Catch basin

Inlet

Outlet

erably, but in some cases, even a little tractor like this is tough to maneuver. Therefore, drainage work on steep hillside tends to be labor-intensive and impossible without a conscientious crew.

Putting It in the Ground

Naturally it's best to think about controlling subsurface water in the planning phases of a new construction project. The pipes can be installed alongside the new footings before the trenches are backfilled. But I can assure you, plenty of houses have been built with inadequate systems for controlling ground water, if any at all. The methods for assembling a system during new construction are the same as that for a retrofit—it's just a lot easier to do it before the foundation trenches are backfilled and the landscaping has taken root. We find that it costs two to three times more to install a drainage system around an established house than it does to add drainage around a new house. The photos illustrating this article show a couple of typical retrofit installations.

Begin with a Trench

When we can get the tractor into place, we begin a job by trenching alongside the foundation (see the top photo on p. 210) until we've reached the base of the footing without undermining it. In the project shown in the photos on p. 210, the house was cut into the hillside to create space for a garage, and the cast-concrete foundation and the stairs to the side yard were gradually being shoved east by the swelling of the clay-laden soils. When this house was built in the '20s, the builders had included foundation drain lines. But they were cast-iron pipes installed a foot or so below grade. When we found them, they were rusted out, clogged with mud, and totally useless. We also found lines as we trenched along the back of the house (see the bottom left photo on p. 210)

and hit water at about 5 ft.—4 ft. above the level of the garage slab.

USE FILTER FABRIC TO PREVENT CLOGGED PIPES

Before we laid pipe in our trenches, we lined the trenches with geotextile fabric. Also known as filter fabric, this material is made of either woven or spun-bonded polyester or polypropylene fibers. The purpose of the fabric is to keep the fines in the soil from migrating into the gravel backfill, which eventually would clog the drain line. We use a 4-oz.-per-yd., spun-bonded fabric designed for soils that don't have a lot of silt in them. Under some conditions, soils that have high sand and silt contents can clog filter fabrics, so if you're in doubt about the makeup of the soil, have a soils lab test its constituents so you can choose a fabric accordingly. Filter fabric is remarkably tough stuff. To cut it, we use sharp sheet-metal shears, but a sharp razor knife will work. We pay about 15¢ per sq. ft. for the fabric and buy it from a local vendor that supplies products related to concrete work. If you can't find filter fabric locally, two companies that make it are Mirafi and Hoechst Celanese Corp.

Before the fabric is down, we sometimes add a thin layer of sand to smooth out the bottom of the trench. This keeps the pipes from getting flattened at the high spots when the gravel backfill is placed. But if the bottom of the trench is pretty uniform, we skip the sand.

LAY THE PIPES

The pipes have to be sloped at a minimum of ⅛ in. per ft. For shallow trenches that have relatively short runs—say 40 ft.—we'll use a 4-ft. level with gradations on the bubble that read slopes of ⅛ in., ¼ in., and ⅜ in. For longer runs, or deeper trenches, we use a transit and a rod to check the slope.

We cut pipes with a hacksaw, and for the most part, we rely on press-fitting the parts because they usually go together with a satisfying snugness. If so, we don't bother gluing them. But if they seem loose, we

It costs two to three times more to install a drainage system around an established house than it does to add drainage around a new house.

Retrofit drainage work begins with excavation of a 12-in.- to 18-in.-wide trench to the base of the footing.

You know you've got a subsurface water problem when your basement is 9 ft. below grade, water is 5 ft. below grade, and the remains of your drain line are 1 ft. below grade. It can be seen in the center of the photo, on the trench's right bank (below).

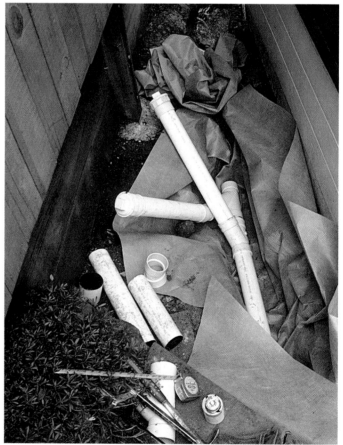

At the high end of the system, cleanouts should be installed allowing access in both directions. Shown above is the subsurface line—the surface line will also need cleanouts.

swab them with the glue supplied by our vendor for the particular kind of pipe and wrap them with duct tape as a further hedge against separation during backfilling.

USE CLEAN GRAVEL TO FILL THE TRENCH

When all the subsurface lines are in place and their cleanouts have been extended above grade, we fill the trench with gravel to within a foot of grade. This gravel should be clean ¾-in. material. If you are applying polyethylene sheeting to the foundation as a moisture barrier, use rounded rock. Crushed rock has sharp edges that will damage the poly. Otherwise, crushed rock is usable and probably cheaper. Don't use road-bed mix, though, because it has too many fines in it. Gravel in place, we wrap the fabric over the top like a big burrito.

Atop all this goes the unperforated surface runoff lines (see the photo on p. 207). While we usually position them about a foot below grade, they can be placed lower, if necessary, for positive drainage or to protect them from the gardener's shovel. The surface runoff lines, too, are sloped at least ⅛ in. to a foot. Leaders from the rain gutters are connected to the surface lines by way of plastic fittings that are square on one end to accept the leader, and round on the other. At each leader entry, we place a wye fitting for a cleanout, and at the highest elevation of the surface drain line we position a pair of cleanouts (see the bottom right photo on the facing page).

Usually the surface line and the subsurface line will drain to daylight at different places, but we sometimes combine the two if we need to go under a sidewalk with a line. In this case we make sure the intersection is at least 20 ft. downhill from the structure to minimize the chance of a blockage that would cause water to back up into the subsurface line. And we include a cleanout.

FINISH THE JOB

To finish the job, we will either compact a layer of soil on top of the buried lines, or cover them with more gravel if the surface is likely to get heavy runoff. Road base is usable here. It's probably overkill, but I think it's best to top the system with some type of paving—either a poured-concrete cap or individual pavers that direct the water away from the building.

What to Do with the Water

The final phase in drainage work is doing something with all the water once you've got a system for collecting and rerouting it. Two guidelines are important to follow. First, if bad drainage is causing your property to deteriorate, then it's important to make sure that your depository doesn't cause the same problems, albeit in a different location. Second, don't put your storm drainage on your neighbor's property. In the latter case when you're on a hill, sometimes the only appropriate solution is to secure an easement for a drain line that will discharge below both properties.

CONNECT DRAINAGE TO AN ESTABLISHED SEWER SYSTEM

In many places storm drainage must be put into established sewer systems dedicated to carrying runoff. Local jurisdictions vary on the hookups required. In some cities you simply daylight the drain lines at the curb sending the water to the storm sewers via the gutters, while other jurisdictions require a sealed hookup. Some will let you dump runoff into effluent sewer lines, but I've found this to be the exception. Given the differing approaches, make sure you check with your local building department to verify local practices.

If you're unsure about the stability of the soils at a likely water-dump site, you should consult a soils engineer.

Options for Runoff Drainage

ENERGY-DISSIPATION BASIN

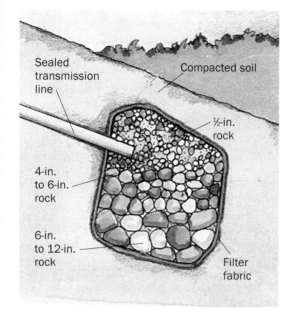

Sealed transmission line

Compacted soil

½-in. rock

4-in. to 6-in. rock

6-in. to 12-in. rock

Filter fabric

DRY WELL

Minimum 1 ft.

Heavy, tight-fitting lid

Transmission line

Filter fabric

¾-in. gravel backfill

Perforated galvanized steel pipe

SUMP PUMP

Subsurface water-transmission line

Cast-iron grate

Basement slab

Moisture barrier

2 ft. to 3 ft

Sump pump

Filter fabric

Washed ¾-in. gravel

OTHER OPTIONS

If there isn't a municipal storm sewer to carry away the water that is collected, things get more complicated. Leach fields, energy-dissipation basins, and dry wells are three approaches to allowing the runoff to continue draining slowly, in a manner less likely to cause erosion and other related problems. But these strategies can concentrate an abnormal amount of water in one place. So if you're unsure about the stability of the soils at a likely water-dump site, you should consult a soils engineer.

If you have an open area with suitable soil, a leach field can be used to distribute the water back into the ground. Like a septic leach field, this requires a manifold of perforated pipes buried below grade. You run a sealed transmission line into one end, and the water will dribble back into the soil over a large area.

An energy-dissipation basin takes less space than a leach field. It usually consists of an excavation lined with filter fabric and filled with rock graded by size (see the top left drawing on the facing page). The ones that we've done have been about 5 ft. square and 4 ft. deep. The graded rock is arranged so that the big ones are on the bottom where they can get a good bite into the hillside. A sealed line enters the basin, and its end should be buried in about 2 cu. ft. of ½-in. rock. Recently we had to convey a load of 4-in.- to 12-in.-dia. rocks 200 ft. down a hillside. We made staging areas at 60-ft. intervals and rolled rocks to them through taped-together Sonotubes®. The crew loved it.

Covered with a layer of native soil, an energy-dissipation basin can be made that will virtually blend into the landscape. But make sure you compact the soil before landscaping to avoid the inevitable settling that will occur.

If there is insufficient grade for positive drainage away from the building, a dry well (called a stand pipe in the midwest) could be your solution (see the top right drawing on the facing page). It is usually placed 15 ft. to 20 ft. from the house. When constructing one of these, we use an 18-in.- to 24-in.-dia. corrugated and perforated galvanized-steel pipe set vertically into an excavation roughly 4 ft. in dia. and 8 ft. to 10 ft. deep. The hole is lined with filter fabric, the pipe goes in the middle, and a collar of ¾-in. drain rock fills the space between them. A sealed transmission line from the drainage system sloped at ¼ in. per ft. enters the dry well. It should be at least 1 ft. deep to protect it from shovels and rototillers.

A dry well should be capped with a heavy, tight-fitting lid that is impervious to youngsters. I hasten to add, however, that dry wells are not appropriate alternatives in all situations. Installed on slopes, they may concentrate water where you don't want it, and could even create an unstable slope condition that might result in a landslide.

As a last resort, you can use a sump pump (see the bottom drawing on the facing page) to lift water out of an undrainable situation. The sump pump goes in the deepest part of the basement and needs a reliable source of power. Two essential characteristics make sump-pump systems less than ideal solutions. For one, if they are installed within the home's footprint, water is still getting in or under the building. And two, they rely on manufactured energy vulnerable to outages that typically occur precisely when you need the pump the most. But in some cases, a sump pump is the only way to get the water to an acceptable distribution system. Regardless of which system you elect to install, it's a recommended practice to keep good records showing the details (size, depth, location, cleanouts) of the system.

David Benaroya Helfant, Ph.D.,M.ASCE, *is managing officer of Bay Area Structural, Inc., general engineering contractors in Oakland, California, and is principal of SEISCO Engineering and Environmental Design Associates, Inc., Emeryville, California.*

Sources

ABT, Inc.®
P.O. Box 837
259 Murdock Rd.
Troutman, NC 28166
800-438-6057
www.abtdrains.com

Heyday Books
2054 University Ave.
#400
Berkley, CA 94704
510-549-3564
www.heydaybooks.com

Hoechst Celanese Corp.
2300 Archfield
Charlotte, NC 28210
www.celaneseacetate.com

Mirafi
A Division of Nicolon Corp.
365 S. Holland Dr.
Pendergrass, GA 30567
706-693-2226
www.tcnicolon.com

Mountain Press Publishing
1301 S. 3rd St. SW
Missoula, MT 59806
406-728-1900

When Block Foundations Go Bad

■ BY DONALD V. COHEN

When people ask me what kind of house foundation I prefer—poured concrete or concrete block—I tell them to take a look at the Yellow Pages under "Waterproofing Contractors." Here in southeast Wisconsin, where I work as a building inspector and engineer, such a search will turn up more than 50 companies specializing in repairing cracks and stopping water seepage in concrete-block foundations.

This is not to say that concrete-block foundations are always a bad idea. They can perform well with proper drainage and appropriate reinforcement, but these conditions are not always present. Gradually, time and the elements can undermine the health of a block foundation, even a well-built one, a fact I am reminded of when doing inspections for prospective home buyers. I constantly see wet basements, or foundation walls that have cracked, buckled, tipped, and sometimes even collapsed (see the photo on the facing page).

Soil Pressure Works against the Foundation

Most problems associated with concrete-block foundations can be traced to two related factors: improper drainage and the seasonal expansion and contraction of soil, which puts pressure on foundation walls (see the drawing on p. 216).

Water seepage is the most common problem I see, but water-soaked soil around the foundation also imperils the structural integrity of the walls. When the ground freezes and thaws, pressure builds against the walls. Common failures are horizontal cracks along mortar joints where the wall has been forced in due to soil pressure. I usually find cracks like this between the third, fourth, or fifth courses from the top in a typical 10- or 11-course wall, which corresponds to the frost line. Often accompanying these cracks are other signs of failure: vertical shear cracks in the corners, step cracks following the mortar joints, and walls

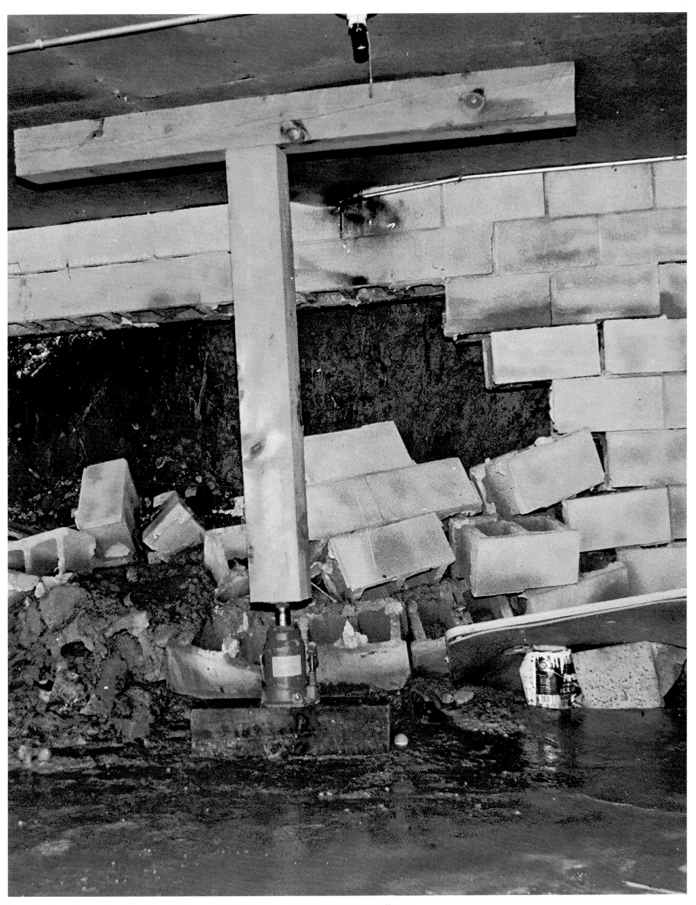

Poor drainage can lead to foundation walls that crack, buckle, or even collapse.

Soil Pressure Can Push a Block Wall off Its Footings

In northern climates, cracks near the middle of the wall are often caused by soil expansion and contraction near the frost line.

Concrete slab

Concrete footing

No mudsill or other lateral restraint

Without lateral restraint at the top of the foundation, pressure can tip the wall inward, sliding it under the joists.

Poor drainage near the footing can lead to soil pressure that forces the wall off its mortar bed between the first and second courses.

Shear crack

Without proper drainage, water gathers in the soil surrounding foundation walls, often finding its way into the basement. Sometimes disaster is unavoidable, as in the photo on p. 215, shot after a storm dumped 6 in. of rain in three hours. Other times, pressure builds slowly, particularly during freeze/thaw cycles. The drawing above illustrates how soil pressure can tip foundation walls inward, cracking the mortar joints. Any wall displaced more than 1 in. from a plumb position must be excavated. Walls displaced less than 1 in. can be braced from inside.

pushed off the mortar joint between the first and second courses of block. Unrestrained walls sometimes slide under the joists in response to soil pressure, tipping the wall out of plumb (see the drawing above).

Water and soil cause other problems, too. Consolidation or settling of subsoil due to heavy rains, or a substantial loss of moisture in the soil, can undermine the foundation from below. This settlement may allow footings to drop, causing vertical and step cracks as well as tipped and cracked concrete

floors. In some cases, this kind of settlement causes the walls to tip outward.

Look for Problems outside the Foundation

The first approach to fixing a wet basement is to correct the grades around the foundation so that water flows away from the walls. Make sure downspout drains, sump-pump discharge pipes, and storm sewers convey roof and surface water away from

Common Failures in Concrete-Block Foundations

Water accumulation and the seasonal expansion and contraction of the soil put lateral pressure on foundation walls. Without proper drainage, the pressure against a concrete-block foundation wall can create failures along mortar joints and cracks in the blocks themselves.

CRACKED WALL

Vertical shear crack

Step crack

Horizontal crack

BULGED WALL

Soil expansion forces blocks in.

BOWED WALL

Gaps appear where middle courses are bowed inward.

the house. Low spots around the foundation are likely to collect water, so any depressions should be filled.

A wet basement may have a deeper source: the absence or failure of a foundation drainage system. For older homes with chronic water problems and no drainage system, the only solution may be to excavate the walls and install a perimeter drain. Clues that might indicate the absence of a drainage system are wet walls, lack of a sump pump, and no evidence of interior drainage or exterior discharge pipe.

Most foundations today, however, are required to have a drainage system that channels water away from foundation walls.

(Residential foundation-drainage systems are generally required by code if seasonal groundwater levels are less than 6 ft. below the surface.) The drainage system—usually concrete tiles, perforated plastic pipe or tubing 3 in. to 4 in. in dia.—is placed along the perimeter footings and covered with crushed stone. The outlet of the drain system can be external (run to daylight somewhere on the lot) or internal (to a sump pump inside the basement). Many modern drainage systems use cross bleeders through the footings from exterior drainage pipes to interior perimeter drains underneath the concrete slab. The cross bleeders are usually spaced 8 ft. apart, starting 4 ft. in from a corner.

If You Catch a Problem Early, You Can Brace Walls without Digging

Walls that are cracked but are displaced less than 1 in. can be braced against further movement. Two common bracing techniques are filling the cores with rebar and concrete, and fastening steel columns to the floor joists and slab.

Reinforcing a wall with concrete and rebar. In this repair, rebar is threaded into the block cores, which are filled with concrete. A ledger strip nailed to the underside of the joists prevents the top of the wall from moving.

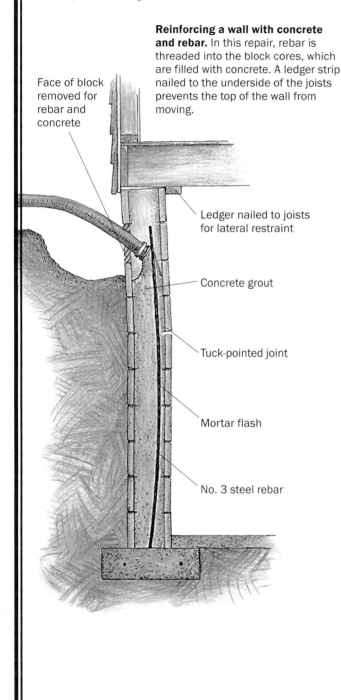

Face of block removed for rebar and concrete

Ledger nailed to joists for lateral restraint

Concrete grout

Tuck-pointed joint

Mortar flash

No. 3 steel rebar

Steel columns provide bracing on the inside. Tubular-steel columns (spaced 32 in. to 48 in. o. c.) are bolted to the joists and the slab to hold a cracked wall in check.

Top of column bolted to joist

Mortar

2-in. by 3-in. tubular-steel columns

Joist

½-in. bolts

Steel plate or washers

Welded base plate bolted to slab

I-joists take braces instead of bolts. Wood I-joists require 2x8 blocking fastened to the underside of the I-joists to lock the steel columns in place.

Steel column

I-joist

2x8 fastened to I-joist with five wood screws

Block wall

If the joists are parallel to the foundation wall, the blocking must be braced as shown below. I-joists require the addition of diagonal bracing.

Diagonal 2x4 bracing for I-joists

Rim joist

Three screws in each joist

Steel column

2x8 blocking notched around steel column

If you suspect a blockage in the drainage system (wet walls along the lower courses, for example), the repair can be as simple as breaking open the concrete floor in the corners and at the midpoints of each wall and flushing the drain tiles. If that doesn't work, you may have to cut open the floor around the perimeter of the basement and replace all the drain tiles and flush all the cross bleeders. If water appears to be trapped inside the concrete-block core, drill a 1-in. hole through the face of the block along the top of the footing to allow any trapped water to seep into the opened drain tile. After the repair, cover the exposed tile with a 1-in.-deep layer of stone, and replace the concrete level with the floor.

Bracing with Concrete and Rebar

In addition to leakage problems, concrete-block foundation walls often display cracks, the early signs of failure. If the walls are fairly dry and if they have not been displaced more than 1 in. from plumb, it is possible to brace the walls without having to excavate the foundation. (The bracing must be designed by a licensed engineer.)

There are two common bracing systems: filling the core of the block with rebar and concrete, and installing steel tubing vertically from the floor joists to the concrete slab. Both methods help the wall to resist soil pressure, sometimes (but not always) checking further movement.

The first method—filling the core with rebar and concrete—requires removing the exterior faces of concrete blocks near the grade line (not at the top), inserting #3 steel rebar into the block cores, and pushing it down to the footing (see the drawing on the facing page). This can be done in every core or as much as 48 in. apart (I have seen it done both ways). The cores are then filled with a concrete slurry mix.

Keep in mind that this repair will not restrain the wall at the top and the bottom.

The concrete slab—if it was poured without expansion felt—will provide support at the bottom. In new construction, a mudsill anchored to the wall (require by code) will provide lateral support at the top. But if there is no mudsill, as in many older homes, a ledger must be installed next to the wall and nailed into the underside of the joists.

Filling a block wall with concrete and rebar is fairly simple, but it has some drawbacks. The concrete tends to hang up on mortar flash inside the block cores, leaving voids in the wall. Another problem is threading 6-ft.-long rebar into the core through the open face of a block. I have seen workers cut the bar and drop the short pieces into the core as the concrete goes in, a method that compromises the strength of the repair.

Filling the core with concrete also causes problems if the wall fails again (if too few cores were filled, for example). A second failure may result in cracked or broken blocks, making it impossible to push the wall plumb. Sometimes, a repaired wall has to be excavated and rebuilt.

Steel Columns Reinforce Walls from Within

Another way to brace cracked walls is to install steel columns inside the basement that span the wall vertically between the footings and floor joists (see the drawing on the facing page). This method is generally a cheaper alternative but does have one drawback: The columns will interrupt any smooth stretches of wall.

The steel columns should be at least 2 in. wide, 3 in. deep, with a $\frac{3}{16}$-in. wall thickness and can be 32 in. or 48 in. apart for a standard 11-course basement. If joists run perpendicular to the foundation wall, the column is fastened to one side of the joist with $\frac{1}{2}$-in. bolts through a steel plate or washers on the other side of the joist. (Wood I-joists require 2x8 blocking fastened to the

> *Concrete-block foundations can perform well with proper drainage and appropriate reinforcement.*

A Reinforced-Concrete Grade Beam Braces Walls after Excavation

For severely displaced walls, a Milwaukee-area contractor has developed—and patented—an alternative to traditional concrete-block foundation repairs. The repair involves excavating the foundation, inserting steel rebar into the core of the wall, and attaching the rebar to a horizontal, reinforced-concrete grade beam. The grade beam is designed to brace a damaged wall to prevent failure in the future.

The process is straightforward: After the wall is excavated and pushed back to plumb, exterior faces in the fourth course from the top are removed at 4-ft. intervals along the length of the wall. The bottom block below the top opening is also opened, and steel rebar is worked up and down inside the core to clear mortar flash at the joints. Debris can be removed from the opening in the bottom block.

With the block cores cleared of obstructions, steel rebar is placed in the cores. The rebar projects out of the block at the top opening perpendicular to the wall, where it intersects a reinforced-concrete grade beam running parallel to the wall. The cores are then filled with concrete after which the foundation wall can be tuck-pointed and waterproofed, and new drain tile can be added.

Crushed stone is then added to within 6 in. of the projecting rebar. Then the 12-in.-high concrete grade beam is formed on top of the stone fill, sloped away from the foundation wall. (The width of the beam can be determined by an engineer but is never less than 12 in.) Soil fill can be placed over the beam to restore the surface to its original condition.

Although this is an extreme measure, the cost of this repair is competitive with other bracing methods. The benefits of a concrete grade beam are its strength and the fact that it does not disrupt the walls inside the foundation.

PATENTED GRADE-BEAM REPAIR
A displaced wall that has been excavated and repaired must be reinforced against failure. One method is to excavate and install a concrete grade beam. The beam is connected to concrete and rebar placed inside the block cores.

Soil backfill

Exterior reinforced-concrete beam

Steel rebar spaced every 4 ft.

Stone backfill

New drainpipe

Asphalt cement

Concrete grout

Block faces removed

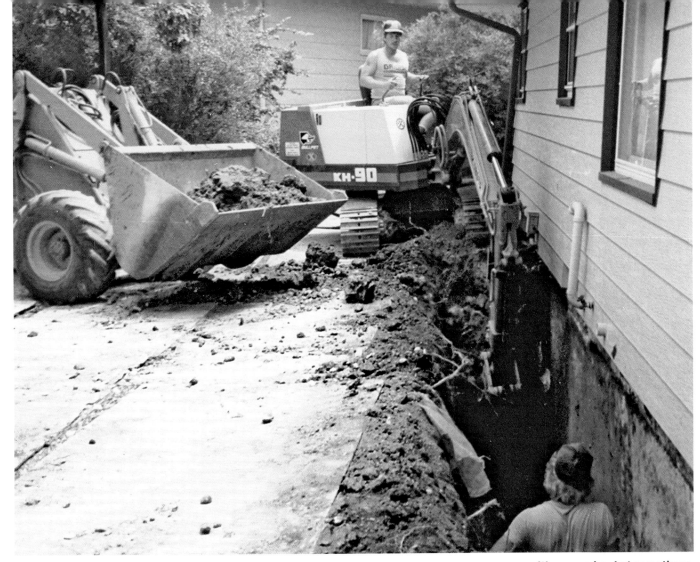

It's expensive, but sometimes excavating the foundation is the only way to fix chronic water problems and severely damaged walls.

underside of the I-joist to lock the column in place.)

If the joists run parallel to the foundation wall, the columns can be held in place with 2x8 blocking notched around the column and pressed firmly to the wall. The blocking should span at least two joist bays to provide lateral restraint. Again, wood I-joists require additional bracing (see the drawing on p. 218).

For lateral support at the bottom of the wall, each column should have a welded-steel plate, which is bolted into the concrete slab or footing. Any spaces between the wall and column can be filled with a mortar grout. If there are pipes or conduit on the walls, they can be accommodated by notching the face of the column to fit over the obstruction, although the notch should be limited to a depth of 1 in. Ductwork along the basement ceiling can be bypassed by weld-ing a horizontal leg to the top of the column at the wall, and then a vertical piece to fit alongside the joist beyond the duct.

Walls Displaced More than 1 in. Must Be Excavated and Replumbed

If water problems go unchecked, the accompanying seasonal expansion and contraction of soil can wreak havoc on foundation walls, sometimes causing severe displacement. Building-code officials here in Wisconsin require any foundation wall displaced 1 in. or more to be excavated and jacked back to plumb (see the photo above).

After the wall is excavated but before it is jacked or pushed back in place, cracked

When Block Foundations Go Bad   221

If Excavation Is the Last Resort

After the walls are re-plumbed and braced, and before backfilling, the exterior of the wall should be coated with fibrated asphalt cement, then covered with 6-mil. plastic sheeting. Drainage also should be added at the footing if none exists.

Topsoil should slope away from the foundation.

Geotextile fabric

Wall is patched, coated with asphalt cement, and covered with 6-mil. plastic sheeting.

Stone backfill

Cross bleeders to interior drain (new construction)

Perforated drainpipe

joints must be cleared of mortar to release the joint and to straighten the wall. With the wall replumbed, the joints must be filled with new mortar. It's important to remember that excavating a finished wall is not enough. Any wall that has been excavated and repaired must be braced or reinforced to prevent future failure. Once the excavated wall has been repaired and before it is back-filled, it's also a good idea to add a drainage system or to repair the existing one (see the drawing above).

Excavation can be an expensive repair. If there is a judgment call as to whether a wall needs to be excavated, as an engineer my decision depends on the extent of the cracking, the amount of moisture in the wall, the distribution of the building load, and the character of the subsoil. Old brick-masonry walls 12 in. thick can bulge inward 2 in. and support a building. Concrete block, on the other hand, tends to fail under similar conditions.

In cases where the wall was constructed out of plumb, a displacement of more than 1 in. might be acceptable, as long as there are no stress cracks. If the same wall is wet, however, excavation might be necessary to repair a drainage problem that might cause a more serious failure in the future.

Donald V. Cohen is a professional engineer, building inspector, and consultant in Milwaukee, Wisconsin. He also teaches courses on home building and structural inspection.

Retrofitting a Foundation

■ BY WILLIAM ANTHONY

It was a cold, snowy day in February when Jack Venning, a second-generation house mover, and I met to check out the project and site conditions. As the cold, wet wind blew off the lake and down the necks of our jackets, it was obvious that winter was not the time to be lifting this house in the air.

The project was the renovation of a cottage that had been sitting on an odd assortment of deteriorating wood and block piers. My clients owned the cottage but leased the land, a common lakeside arrangement here in western New England. Tearing down the cottage initially would have left them with nothing on the land and with no legal claim to the site, and getting new construction authorized would have been nearly impossible. Also, because of local regulations, leaving the cottage in place meant fewer restrictions for installing an urgently needed new septic system. The solution seemed simple. We had to keep the existing cottage and build a new foundation for it. That's where Jack's expertise would come in.

No Place to Put the House

Because the leased lot was so cramped, we had no room to move the cottage away from its footprint. The only way to go was up. The building had to be lifted high enough to excavate for the footings, foundation, and crawl space. But to complicate matters, the cribbing that would support the house during all this work had to be well inside the perimeter of the footprint so that we could dig our frost-wall trenches.

The steel beams that the house was to rest on during its time in midair had to be beefy enough to cantilever from the cribbing piles out beyond the perimeter of the house. The beams and cribs also had to be strategically placed to leave space under the house for the piers and LVL beams that would support the floor of the rebuilt house.

Before the foundation work can begin, house movers are called in to lift the house out of the way.

Keeping Water Away from the Site

Before we could start the lift, we had to think about water, both rainwater and the water in the lake that was a mere 20 ft. from the back of the house. Working with an engineer's approved plan, we staked hay bales and siltation fencing along the water's edge as an erosion-control barrier.

The fencing, available at most contractor-supply houses, comes with the stakes already attached to hold it upright. The hay bales were staked in place with the cut edge down so that the bales could follow the contours of the ground. We also built a 20-ft.-dia. siltation pond of hay bales and fabric fencing. Any water that got into the footing trenches would be pumped into our pond where the silt could settle out before the water seeped back into the lake.

We also added some runoff-control measures of our own. For very little money, we installed gutters and downspouts on the cottage that were then connected to black-plastic flex pipe to channel roof water away from the excavated foundation hole. On the uphill side of the cottage, dirt berms and secondary hay-bale barriers helped to redirect runoff from above the cottage. With erosion-control measures in place and approved, we were ready to begin the lift.

Elevation and Levitation

Our little cottage had seen seven distinct construction phases in its history. Like many area cottages, this one probably started as a summer tent platform on concrete blocks. According to some of the locals, the first structures were the remains of trolley cars from an abandoned local line. A curious rectangular kitchen with a curved ceiling suggested that our cottage may have begun as one of those trolley cars.

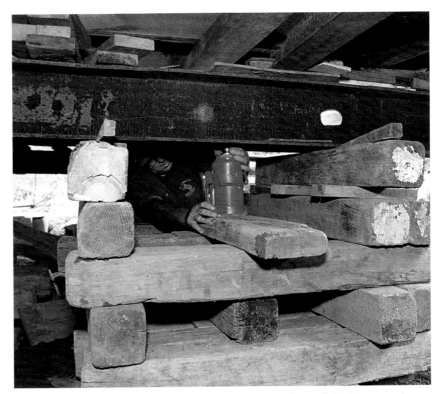

First, steel beams are placed strategically under the floor of the house, and cribs made of stacked-up timbers support the jacks that do the raising.

Measurements are taken to keep the house rising evenly.

All these additions meant that the house would act like an accordion if not supported evenly during the lift. To counteract this effect, Jack spanned the width of the cottage with five 30-ft. steel I-beams, each rusted and scarred from years of house-lifting. Each beam was supported by cribbing piles set in 5 ft. from the perimeter. The cribbing piles, or cribs, served as the platforms for the beams and for the 30-ton hydraulic jacks that did the lifting.

Limited space on the lakeside of the house meant that the steel beams had to be inserted, and eventually removed from, the uphill side. But from years of settling, the cottage had limited access on the uphill side. To make room for the cribbing and beams, Jack first nudged the house up with special ratchet jacks until the crew had room in the crawl space to shovel out spots for the cribbing piles.

The beams were then slid under the house, and Jack's crew put solid blocking at each point where the beams contacted the underside of exterior sills. Additional blocking and smaller secondary steel beams under the house picked up the loads of the interior bearing walls.

Some projects are lifted with a series of interconnected hydraulic jacks that can be operated all at once by a small crew. But for this project, each beam had its own separate jack. Each crew member was responsible for a beam, alternating between the cribs at one side of the house and the other (see the photo on p. 224). The cottage was lifted and blocked about 3½ in. at a time until one side was high enough to insert a layer of 6x6 cribbing (see the top photo on p. 225).

The jacks sat in the center on slightly thinner cribbing pieces inside the crib that could be easily slipped out and reset at the next level as the cribbing piles were increased one layer at a time. The lifting process went on for several days with Jack measuring the height after each increment (see the bottom photo on p. 225). We stopped at a height of about 8 ft., which was enough to allow a small excavation machine space to work under the house safely.

Excavating with Low Overhead

I have to admit that it was more than just a little unsettling watching a Bobcat® loader dodging and weaving between the cribs with tons of house hovering above (see the photo on the facing page). But my excavator, Bob Smith, and his crew did a masterful job of carefully removing all the material to grade from between the cribs.

The excavation left each crib on an undisturbed mound of earth ranging from 30 in. high on the uphill side to 6 in. on the lakeside. Soil tests that had been done earlier had shown that the glacially compacted soil that was under the house could carry the cribs safely with near-vertical excavation.

Working on a tiny lot meant that no excess material could be stored on site. Instead, the loader brought all the extra soil over to the corner of the excavation nearest the road (see the top photo on p. 228). An excavator then scooped the soil out of the hole and onto a waiting truck. Many truckloads of material were carried to the top of the hill behind the cottage and stored until it was needed later for backfilling and grading.

A second Bobcat with a backhoe rig dug the frost-wall trenches (see the bottom photos on p. 228). The soil under the cribs was stable enough to let us dig our trenches without undermining the cribs. We were able to keep our footing grades and trenches to the required depth, which included a 12-in. layer of compacted stone on top of geotextile fabric (see the bottom right photo on p. 228).

The only soil that stayed on site was what we'd need to backfill the trenches after the foundation was poured. These soil piles were kept low to minimize erosion and were tarped. Tarped berm piles along the uphill side acted as a water diverter to keep the excavation sides intact. Due to our antiflood strategies, the small amount of water that seeped into the trenches was handled easily with a small submersible pump.

With tons of house suspended above them, workers in small excavation machines dig their way around the cribs, leveling the floor of the hole.

Extra soil is taken to one corner of the trench where it is scooped up, loaded onto trucks, and hauled away until needed for backfilling.

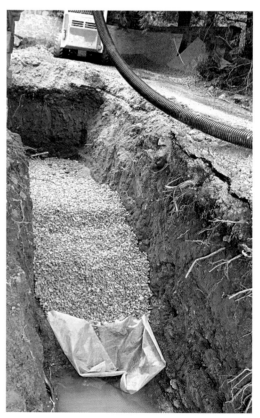

Trenches are dug with a minibackhoe (above), and the bottom of the trench is covered with filter fabric and a layer of crushed stone (right).

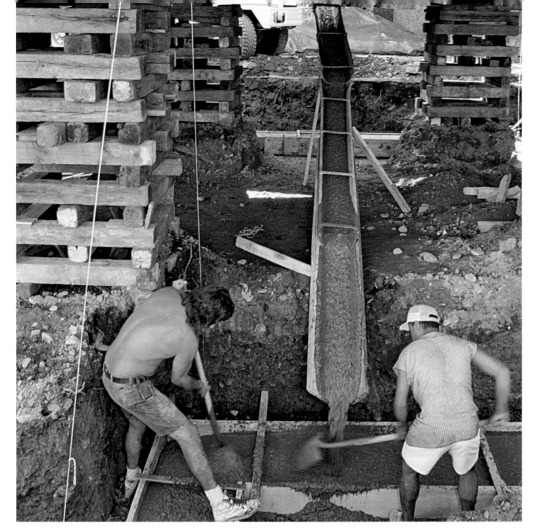

Footings are formed and poured on top of crushed stone with long chutes to make up for poor cement-truck access.

A frost wall is poured on top of the footings.

A Little Concrete Goes a Long Way

The footings on the front and rear of the cottage were on one level and formed with standard 2x forms. The side footings stepped down grade and needed elaborate formwork.

Access for the cement trucks at the site was limited, so my concrete contractor, Bob Kallenbach, rigged up a 30-ft. chute to get the mix to the far side of the house (see the photo above). Next, the concrete side-walls were formed and poured high enough so that the steel beams could be removed after the house was lowered (see the photo at right). We attached small site-built forms inside the main forms to create a 5-in. shelf for stone veneer that would be added later.

Before the foundation was backfilled, footing drains and gutter drains were installed. The frost walls were waterproofed, and insulation/drainage board was applied

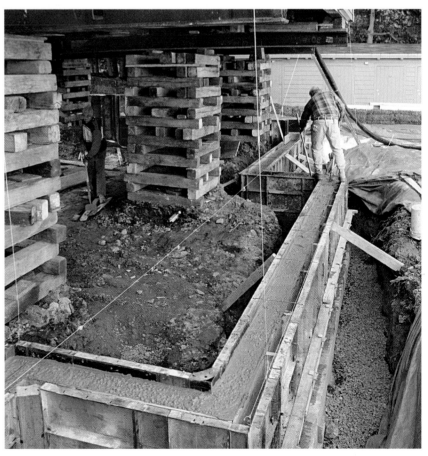

Site-built forms create a shelf for the stone veneer. Insulation board covers the frost wall before backfilling.

A block wall is then laid with pockets left for the steel beams.

to combat the water table (see the top photo at left). Then concrete block was laid on top of the foundation walls to make up the rest of the crawl space–wall height with holes for beam removal (see the bottom photo at left).

Bringing Down the House

The foundation walls on the uphill side of the cottage created a 5-ft.-high crawl space. But where the grade dropped toward the back of the house, we stick-framed the crawl space walls. Breaks were left in the top plates so that the steel beams could drop below the top of the wall. The breaks were filled in after the house was lowered.

The day came to lower the cottage onto the walls we'd built (see the top photo on the facing page). The house was lowered in 3½-in. increments, the reverse of the lifting process. It's strange when all five crew members open the valves on their jacks at the same time and, with a long hiss and lots of creaking, the building slowly settles.

It took about a day and a half to lower the house and level it on the crawl space walls. My mason, Pat Varrone, then built the block piers inside to accommodate the LVL replacement beams. When the LVLs were positioned and blocked in place, the weight of the house was taken off the steel beams (see the bottom left photo on the facing page). The beams were then rolled out through the pockets we'd left in the block wall (see the bottom right photo on the facing page).

With the house supported, we used the soil that had been under the cribs to backfill the inside of the frost wall. Before blocking up the beam holes in the foundation, we made use of the access to pour the slab in the crawl space. The holes in the block wall and the wood-frame wall were then filled in, and Pat got to work on the stone veneer.

By the end of the next summer, the project was well on its way. Down came our stout cottage that served so well as a tem-

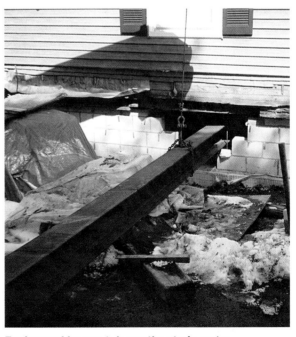

The house mover drops the house back down onto the completed foundation.

plate for its successor. We were able to save large portions of the first-floor deck left as a reminder, like the hint of the old trolley in the kitchen, of previous generations that enjoyed this little place by the lake.

Engineered beams take up the strain under the house (above left), and a crane slips the beams out through the holes in the foundation (above right). Renovation of the house can now begin.

*Before retiring, **William Anthony** of Woo dbury, Connecticut, owned and operated a building company that specializes in weekend and second homes, as well as old-house renovation*

Four Retaining Wall Choices

■ BY ERIC NELSON

A homeowner once approached me as I stood, hammer in hand, working on a stone retaining wall, and asked, "Is that it? You just knock the rocks into place?" I wish it were that simple. Stacking natural stone or concrete pavers or spiking together timbers may look easy, but I've replaced or rebuilt enough walls to know how often it is done incorrectly. The trick is to pick the material that best suits your landscape, your budget, and your ability (if you plan to tackle the project yourself).

I usually build fieldstone or timber walls because they are the most popular materials in my area, but modular-block and poured-concrete (or concrete-block) walls also are excellent options. Each type of wall requires a different set of tools and building skills. For instance, timber landscape ties are a good, inexpensive material for someone comfortable building with wood. A poured-concrete wall is likely the strongest choice, but the installation is probably best left to professionals.

Choosing the most appropriate material is the first decision you need to make when starting a retaining-wall project. Before you begin building a wall, you also need to consider the elevations of the new finished grades, the appropriate base and backfill, and the question of whether an engineer should design the wall.

When Do You Need an Engineer?

You'll have to check with your local building department to find out if you need a permit and engineer's approval to build a retaining wall. The International Residential Code regulates the construction of walls more than 4 ft. tall. Likewise, most manufacturers of modular-block wall systems recommend that any wall over 4 ft. be designed, or at least approved, by a professional engineer. I agree that there is little danger in a homeowner or contractor tackling a project less than 4 ft. tall. I also agree that it is a good idea to speak with an expert before attempting to build a retaining wall over 5 ft., even if your local code doesn't require that you do. In many cases, two small walls look better than one tall wall anyway.

Consider Terracing

Even the most attractive retaining wall can lose its charm if the wall is too massive for its surroundings. If there is enough area to divide the change in grade into multiple stepped walls, a series of small walls might be more visually appealing. Terraced walls also may preclude the need for permits, engineers, and complicated construction details. The area between terraced retaining walls does not have to be level and can be used for lawn or gardens.

Building two or more small walls also may be easier on your back because you don't have to lift the stone, block, or timbers too high, or set up staging. If you decide to build a terraced retaining wall, start with the lowest wall: You'll create flat areas to stand on while working on the upper walls, and you can incorporate steps, which have to be built from the bottom up.

Base and Backfill

One truth about all retaining walls is that they are only as good as the base they are built on. The right depth and type of base depends on the material and the landscape. The base and backfill in the drawings shown here represent best practices. In other words, this is how I would build each of these walls in wet or uncertain conditions. In dry areas, I build timber walls right on the ground.

Stone veneer hides an unattractive retaining wall.

Concrete

A Strong Material with Many Looks

Poured-concrete and concrete-block, or concrete-masonry-unit (CMU), walls are the most complicated, most expensive, and least attractive retaining walls. If they're engineered well, though, they're probably the strongest. And veneers can take the curse off how they look.

These heavy walls always should be designed by a professional, who will test the soil conditions and draw plans for the project. Because they demand extensive excavation for the footing and material to build the forms, the pour is also best left to professionals. CMU walls require a footing engineered similarly to poured-concrete walls, and their strength depends on the right mortar mix and reinforcement.

If you need retaining walls near a house under construction, consider having them poured at the same time as the foundation.

Laying up veneer can be tackled by a first-timer. Although the process is slow and the materials expensive, the wall's structural integrity won't be threatened by mistakes.

There are a variety of veneer materials to choose from. Natural stone is popular, but thin stone veneer is more expensive than most wall stone. Cultured, or artificial, stone is less expensive and has the advantage of uniform thickness. Of course, in the right landscape, brick veneer and stucco are good choices, too.

Vertical and horizontal rebar

Galvanized ties

Filter fabric

Gravel backfill

Weep hole

Perforated pipe

Haunch

Poured footing

Vertical and horizontal rebar

Concrete Walls Take on the Look of the Surface Treatment

Stucco

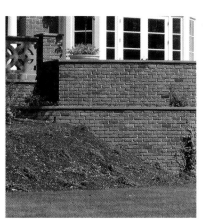

Brick

And although I always build stone and modular-block walls on at least an 18-in. gravel base, I put a drain in the base only in wet areas. Any time a perforated pipe is used in the base, the gravel must be separated from the earth with filter fabric.

Backfill is equally important. Dry-laid stone walls usually are backfilled with large stones, and the voids are filled with rubble. If the wall is separated from the earth with filter fabric, drainage behind the wall often is unnecessary. Modular-block walls should be backfilled with gravel and a perforated drainpipe and separated from the earth with filter fabric. This method also can be used for a timber wall in a wet area, a wet-stacked stone wall, or a poured-concrete wall. Weep holes are another option for poured-concrete walls and wet stone walls.

For some projects, excavating for the base and backfill is the most laborious part of the job. If this is the case, it might be worthwhile to hire an excavator or to rent a backhoe for a day.

The last thing I do before I begin a wall is to determine the height of the top of the wall and the finished grade at the bottom of the wall. This way, I can make sure the top of the base is set a few inches below the finish grade (lawn or garden) to keep it hidden when the project is complete. Don't trust your eye when determining elevations. Use a site level to avoid mistakes.

Eric Nelson owns and operates Garden Paths, a small landscape company that focuses on properties where landscaping complements natural surroundings. Nelson and his family live in Bethlehem, Connecticut, where he has worked as a landscape contractor for 22 years.

One truth about all retaining walls is that they are only as good as the base they are built on.

Sources

Allan Block Corporation
5300 Edina Industrial Blvd., Ste. 100
Edina, MN 55439
952-835-5309
www.allanblock.com

Unilock
35 Commerce Dr.
Uxbridge, MA 01569
800-864-5625
www.unilock.com

Versa-Lok
6348 Highway 36
Ste. 1
Oakdale, MN 55128
800-770-4525
www.versa-lok.com

Keystone Country Manor
4444 W. 78th St.
Minneapolis, MN 55435
800-747-8971
www.keystonewalls.com

Modular Block

Versatile in Design and Interlocked for Strength

If you want the look of a natural-stone wall but lack the skill to build one, a modular-block wall might be the answer. These wall systems come in a variety of styles, patterns, and colors. There are tumbled blocks of uniform size that simulate the look of quarried granite, blocks of varying sizes that form patterns to look like natural stone, and split-faced blocks that look like what they are: concrete. In fact, these blocks are made from really strong concrete; most have a compressive strength of 5,000 psi.

Modular-block retaining-wall systems are available from a number of manufacturers. Prices vary between manufacturers and styles. Each company—sometimes each style—has its own building system. Most systems are comprised of a few different components, including the basic wall blocks, corner blocks, and cap pieces. The good news is that none of the systems is too complicated, and the walls go up quickly.

The trick is to get the first course level. Then it is mostly a matter of stacking blocks and backfilling. Many systems even incorporate a setback into the design so that as you build up, the wall automatically pitches back into the retained earth behind. It is hard to avoid cutting blocks, but the only specialty tools you'll need are a masonry or diamond blade for your circular saw and a mason's chisel.

These interlocking systems are versatile enough for most designs. Curved walls can be built, and matching steps and walks can be incorporated. Some manufacturers will send a representative to help you figure out just what you need for your project.

With Pins or Ridges, Modular Blocks Lock Together for Strength

Unilock® Roman Pisa®.
A ridge on the top of one course fits into a groove on the bottom of the next.

Versa-Lok® Mosaic®.
A top-down pinning system lets you push a pin through a hole in one course and into a groove in the course below.

Keystone® Country Manor.
Pins are placed into the top of one course, and the next course is positioned on the protruding pins.

Caps, or coping blocks, are glued onto the final course.

Gravel backfill

Filter fabric

Stagger the seams.

Perforated drainpipe

Gravel base

Random patterns look natural. Allan Block's® Ashlar Blend is a combination of three different blocks that can be mixed and matched.

Timber

Less Durable, but Easy to Build

For most folks, timber retaining walls are the easiest type to build. And in most cases, they are the least expensive option (8-ft. pressure-treated 6×6s start at $16). If you are comfortable with basic carpentry, you shouldn't have a problem building a timber retaining wall. In a dry area, a timber wall can be built without a base or tricky backfill. Steps can be made by joining two 6x6 timbers and are incorporated easily into these walls. It is simple to make 90-degree and 45-degree corners, though curves are tricky.

Timbers range from used railroad ties to planed lumber. If you can find them, used railroad ties are fairly inexpensive, but they're often treated with creosote. They also may be inconsistent in dimension and not very straight. This combination can make for tricky building.

The most commonly used material is pressure-treated 6x6 landscape timbers. You can use a rough-sawn, dimensional 6x6 for a rustic look, or a planed

6x6 (actually 5½ in.) for a more-finished look. For shorter walls, a 4x6 timber also can be used.

The length of the ties also varies. A 12-ft. or 16-ft. tie can cover a lot of ground but is difficult to handle alone. An 8-ft. tie is more manageable for one person.

The obvious drawback to timber walls is that they eventually rot. Pressure-treated timbers will last longer than untreated timbers, but be aware that the manufacturer's warranty on new pressure-treated timbers may not be honored if the tie is cut.

Filter fabric

Gravel backfill

Deadman

Perforated
drainpipe

Tieback option: If there is
insufficient space for long
deadmen, a tail adds strength.

Staggered
corners

10-in.
spikes

Buried first
course

Gravel base

Rebar

The Basic Steps of Timber Wall Construction

1. Cut timbers with a chainsaw, or with a circular saw and a handsaw.

2. Drill pilot holes so that spikes or rebar can be driven easily, with less chance of splitting.

3. Tie courses together by driving spikes with a 12-lb. hammer or 20-lb. sledgehammer.

Durable, but Construction Is Tricky

Stone walls may be the best choice for a natural-looking landscape. Natural stone offers limitless design possibilities, including curves, and stone walls can be built to follow a sloping grade. Steps can be incorporated into stone walls, though stone treads often are too heavy to handle without a machine. The price, type, and availability of stone vary from one area to the next, but natural stone is almost always one of the more expensive retaining-wall materials.

Along with pricey material costs, stone walls bring high labor costs. It takes practice to learn to build with the irregularities of natural stone, and even for an experienced mason, building a stone wall takes longer than building the same wall with another material. Still, stone walls can be a rewarding project for patient first-timers. In most areas, you can have pallets of stone delivered. A few inexpensive tools, such as a mason's hammer, will make the work go more smoothly.

Stone walls can be stacked dry using stone and rubble for backfill. Dry walls are built on a base of compacted gravel. It's important for hidden backfill stones to be stacked just as securely as visible "face" stones. All voids inside the wall should be filled with rubble.

Another option is to stack the face stones dry, then backfill with stone and mortar. This type of construction requires a deeper (24 in.) compacted-gravel base. Finally, you can mortar the joints between stones. These walls should be built on a poured-concrete footing with rebar placed horizontally in the footing and vertically to extend through the wall as it is built up. All "wet" walls need a drain in the backfill or weep holes to relieve pressure from water that seeps behind the wall.

Stone walls can be deceiving. From a distance, this stone retaining wall appears dry-stacked. A closer look reveals that the stones actually are mortared. Mortared backfill is a good idea for taller walls or walls that will be walked on or sat on.

Three Ways to Stack a Stone Retaining Wall

1. DRY-STACKED WALLS ARE BACKFILLED WITH RUBBLE

Long stones tie front and back together.

Voids are filled with rubble.

Filter fabric

Gravel backfill

At the base, the wall is two-thirds as thick as it is tall.

Flat stones are used for the cap.

Stagger the seams.

Square cornerstones

Perforated drainpipe

18-in. gravel base

Vertical rebar

Mortared joints

Poured footing

Horizontal rebar

3. A POURED-CONCRETE FOOTING PREVENTS WET WALLS FROM CRACKING

The cap stones can be mortared or left dry.

Mortared backfill

24-in. gravel base

The face stones are stacked the same as a dry wall.

Weep holes (1-in. PVC pipe) let water out from behind the wall.

2. THE DRY LOOK, WITH THE STRENGTH OF CEMENT

The many faces of natural stone. The availability and type of stone vary from one area to the next. Fieldstone is easy to find in the Northeast (photo, facing page). Quarried Niagara boulders (photo, below) and limestone (photo, right) are common in the Midwest.

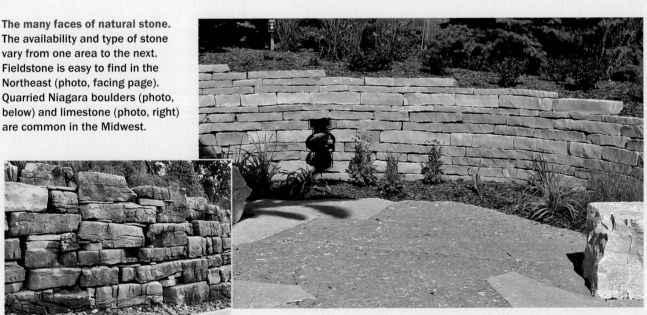

Building a Modular Retaining Wall

■ BY ANDY BEASLEY

When my wife and I had our house built, the budget was far too tight to consider professional landscaping. When finished, the house looked as if it had been dropped on the site of a bomb blast. What was worse, any rain that fell carried silt toward the back door, and on windy days the back of the house was scoured with airborne grit. My wife tactfully suggested that we either needed a retaining wall and a patio to keep all that soil at bay, or we might as well board up the back of the house.

She was right, but what kind of wall? I looked at several options: poured concrete, concrete with stone veneer, timber, dry-stack stone, and concrete block with stucco facing. All of them have advantages, but they were either too expensive or too time consuming to construct, or didn't have the right look. I was about out of ideas when I came across something called a segmental retaining wall: basically, a wall made of interlocking pieces of cast concrete. I liked the way it looked;also, it wasn't too expensive,

and construction seemed simple. Clearly, this stuff was for me.

Several companies make this kind of wall system, and they share a number of similarities (see Sources on p. 251). A segmental wall is made of concrete blocks that stack together without mortar. Once finished, the wall has enough mass to retain the soil behind it. Each course is set back slightly from the one below, so the wall steps back into the bank as it rises. Corners, curves, and stairs can be built, although specially shaped blocks may be required.

Segmental walls up to about 4 ft., like the one I planned, are considered "gravity walls," meaning they need nothing more than the force of gravity to keep them in place. (Higher walls are typically reinforced with a polyester filament geogrid that ties the wall to the compacted earth behind it.)

Concrete blocks may be solid or hollow and use either pins or integral lugs to align the units and lock them together. The finished faces and colors of the block

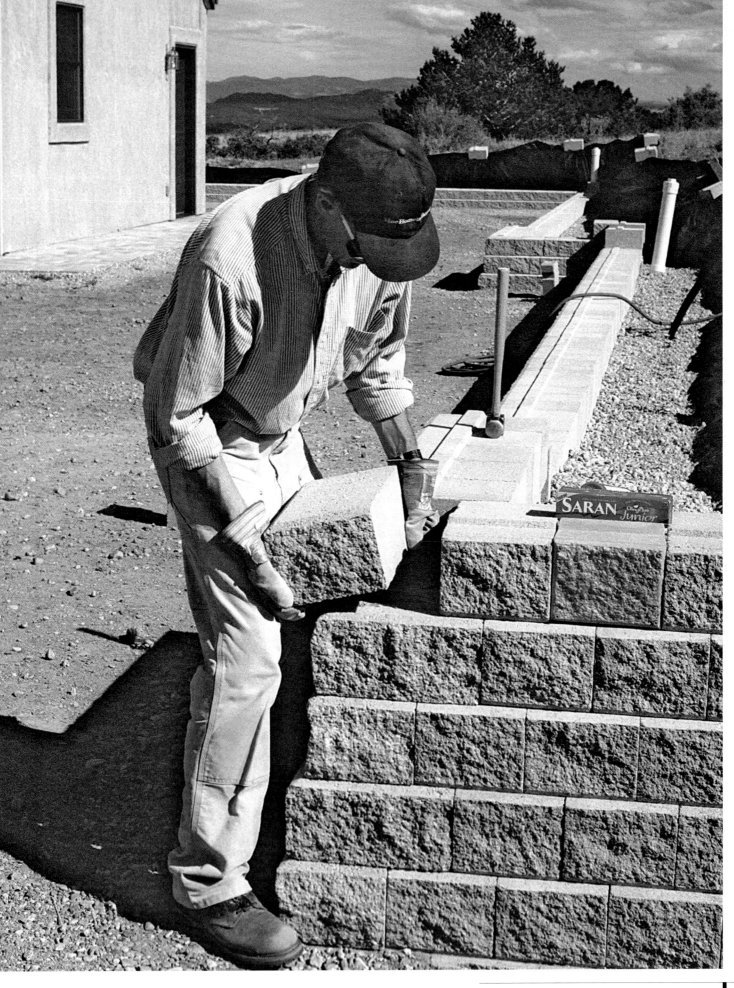

Stepped for Stability

As the wall rises, it also steps back into the bank for more stability. An interlocking tongue-and-groove pattern automatically steps each course of block back from the course below. A layer of geotextile, or filter fabric, prevents silt from packing the crushed stone behind the wall and interfering with water drainage.

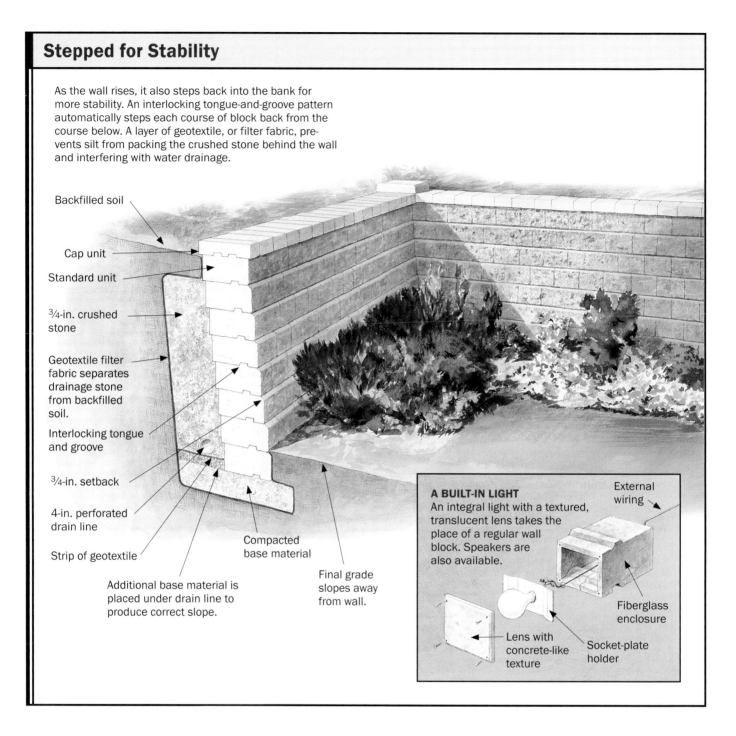

Backfilled soil

Cap unit

Standard unit

³⁄₄-in. crushed stone

Geotextile filter fabric separates drainage stone from backfilled soil.

Interlocking tongue and groove

³⁄₄-in. setback

4-in. perforated drain line

Strip of geotextile

Additional base material is placed under drain line to produce correct slope.

Compacted base material

Final grade slopes away from wall.

A BUILT-IN LIGHT
An integral light with a textured, translucent lens takes the place of a regular wall block. Speakers are also available.

External wiring

Fiberglass enclosure

Lens with concrete-like texture

Socket-plate holder

vary by product line and regional manufacturer, and the weight of the individual blocks can differ significantly. I chose a product called Pisa2® from Risi Stone Systems. The solid blocks have a rough-finished face and are 6 in. high, 8 in. wide, and 12 in. deep, weighing 45 lb. each. Pisa also offered accessories, including a built-in light I liked enough to incorporate into my plans. Just as important, I was sold on the strong reputation of McKinney Concrete, the regional

manufacturer that makes the blocks for Risi. Excluding shipping, blocks and cap pieces cost about $4.50 per sq. ft. of wall face.

Wall Begins with a Good Foundation

My plans called for a total of 220 ft. of wall with a series of jogs and one recessed section in the middle for a pair of large planters.

You may think that instructions are for sissies, but I found Risi's material invaluable in helping me to avoid layout and construction problems that would have been difficult to fix midway through the project. If I had requested it, engineering support also would have been available from Risi.

I began with a trench about 30 in. wide and roughly 12 in. deep. That depth allowed room for 6 in. of compacted base material and one block below finished grade (see the drawing on p. 244). Because the foundation is well above the frost line in my area, I installed drains to carry away water that would freeze and heave the wall. And because the wall is not mortared, it can handle minor seasonal movement. One point to remember before putting away the backhoe is to make sure the excavation is sloped back at least enough to accommodate the lean of the wall. In my case, the wall stepped back ¾ in. for each course, or a total of 5¼ in. over the wall's 4-ft. height.

After compacting the soil in the trench, I laid geotextile material from Mirafi® across the bottom of the trench and up the bank behind the wall. This heavy-duty polyester fabric allows water but not fine aggregates or silt to pass through. It serves two purposes here. First, it isolates the base material that goes in the bottom of the trench from the surrounding soil, which I think will help to keep the foundation layer more cohesive over time. Second, the fabric keeps the crushed drainage stone behind the wall from becoming clogged. On top of the fabric in the trench, I compacted a 6-in. layer of road-base material that was available locally (see the top photo on p. 246). Road-base material is a mixture of fine crushed rock and gravel that compacts well but is still porous enough to allow water to drain through.

To make sure my foundation layer was level, I drove pieces of rebar into the ground and aligned the tops of these pins with a builder's level. Once I brought the base material to the tops of the rebar, I knew the foundation would be level, essential for a successful wall (see the bottom photo on p. 246).

Correcting Mistakes Early Is Easier in the Long Run

Like the best-laid plans, the blocks will sometimes go awry. To keep the wall stable and attractive, the top surface of each course should be level and straight. Errors do creep up, and they can compound into real problems unless they are fixed.

How much error is too much? That's a subjective issue. I usually fixed any misalignment of adjacent blocks of more than $\frac{1}{32}$ in. Generally, correcting errors was easy. My first step was to remove the block and check for interference from gravel or concrete chips underneath. Next, I tried switching blocks, because individual blocks may have slight variations. These two steps corrected the vast majority of problems. If, however, I still wasn't satisfied, I would temporarily stack a few blocks of the next row above the problem area, which sometimes showed the problem was canceled out. If so, I would forget about it and move on.

If these steps were not enough, I'd resort to shims to level the row. Shims can be almost any material, but they should be something that will neither erode nor decompose. I used thin strips of aluminum when necessary. In the extremely rare cases where a shim would only compound the problem, I removed the irregularities by taking light passes with my trusty circular saw with a diamond blade. A grinder would also work.

A solid wall needs a solid base. After digging a 12-in.-deep trench 30 in. wide, the author compacts a layer of locally available road-base material over a layer of filter fabric to serve as the wall's foundation.

A level start is crucial. Rebar pins driven into the ground on 6-ft. centers act as reference points to level the compacted base material.

Get the Base Course Right

If there's one thing I learned from this project and from studying other retaining-wall jobs, it's that getting the first course level is difficult but crucial. Because the block faces are rough-textured, I kept things on course with a string along the back of the units. I placed each block next to the one previously set, tapped it with a hammer to settle it on the base, and leveled it both side to side and front to back.

The blocks rarely lined up perfectly on the first attempt, so it became a process of setting a block, checking to see how it lined up with its neighbor, removing the block to adjust the base material, resetting and rechecking the block, and so on until it was level and exactly aligned. I used a modified torpedo level (see the bottom right photo) to adjust each block front to back and a 6-ft. level to align the block with its neighbors in the row. I sighted along the bottom of the long level and knew a block needed correction when I could see daylight between the level and the block. Any difference greater than $1/32$ in. had to be corrected (see the sidebar on p. 245). That may seem overly fussy, but variations greater than $1/32$ in. meant succeeding rows would not sit as firmly as I liked. And errors—especially in the first course—have a way of magnifying as you move upward.

To form outside corners, I used specially designed blocks that have textured faces on two sides. To form inside corners, I used standard units in alternating directions so that the corners interlocked (see the bottom left photo).

The base course also is the time to think about objects that will be built into the wall later, such as speakers, lights, or water lines (see the bottom right photo on p. 249), which may have to pass through. Planning and adjusting the base course accordingly can prevent headaches farther up the wall.

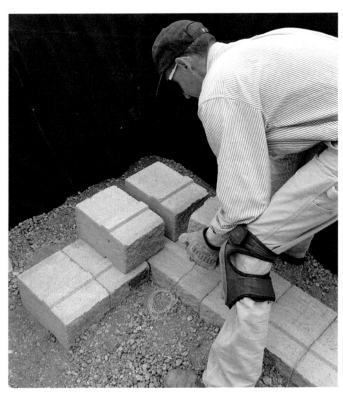

Inside corners are woven. The author sets an inside corner on the first course of block, establishing the spacing for a woven pattern to give the corner interlocking stability.

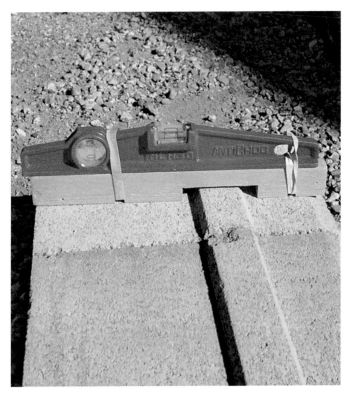

Modified for the job. A simple jig allows this torpedo level to straddle the lug cast into the top of each block so that it can be leveled front to back.

Drains and Clean Aggregate Direct Water Away from the Wall

One of the greatest enemies of any retaining wall is the buildup of hydrostatic pressure from groundwater behind the wall. A well-constructed drainage system is essential to managing the flow of water.

Due to site constraints, I was forced to install fairly long runs of drain, so I took extra time to make them efficient. Starting at the drain outlet, I sloped and compacted ad-ditional road-base material behind the first two courses of block. On top of this slope (about ⅛ in. per ft.), I laid a strip of geotextile the width of the trench to separate the base material from the larger aggregate to come. Next, I put 4-in. perforated drain pipe on top of the fabric. I added clean-out risers with caps every 50 ft. and leaned them backward to ensure that they wouldn't interfere with the slope of the wall (see the left photo on the facing page). After testing the lines, I carefully covered the pipe with clean ¾-in. crushed stone behind the blocks and tamped it in place.

Clean Cuts in Concrete Block

There's no way to avoid cutting block, and I got plenty of practice in the course of building 220 ft. of wall. Renting a block saw or a hand-held concrete saw would have made the work go much faster. But because I didn't work on the wall steadily, it would have cost a small fortune in rental fees to keep one of those saws on hand. Instead, I made all my cuts with a circular saw fitted with a low-cost (around $20) diamond blade. I found that pencil marks quickly disappeared in the cloud of concrete dust, so I used strips of duct tape to guide my cutline around the four sides of the block (see the photo at left). A good respirator, safety glasses, and hearing protection are essential. I cut blocks in three passes to the full 2½-in. depth of the blade (see the center photo). Once I'd cut all the way around the block, I used a 2½-in. masonry chisel to separate the pieces (see the photo at right). I cleaned the excess material from the break at the center of the block by sweeping the sawblade from side to side to make light cuts.

Tape marks the line.

Cut blocks in three passes.

Finish it with a chisel.

First the block, then the stone. After laying a course of block, the author brings up the level of crushed stone behind the wall and tamps it by hand.

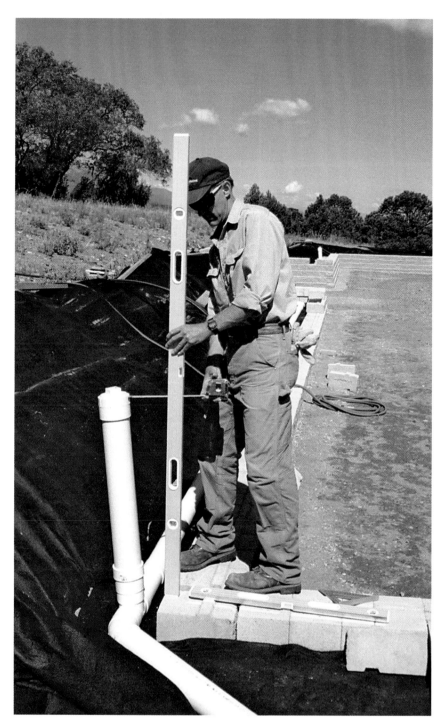

Plumb is not what you want. Clean-out risers for the subsurface drains behind the wall must be pitched back at the same angle as the wall. Over the wall's 4-ft. height, the face tilts back a total of 5¼ in.

Make way for water. A circular saw with a diamond blade makes the narrow channels for these water lines. Grooves are patched with mortar and buried during final grading.

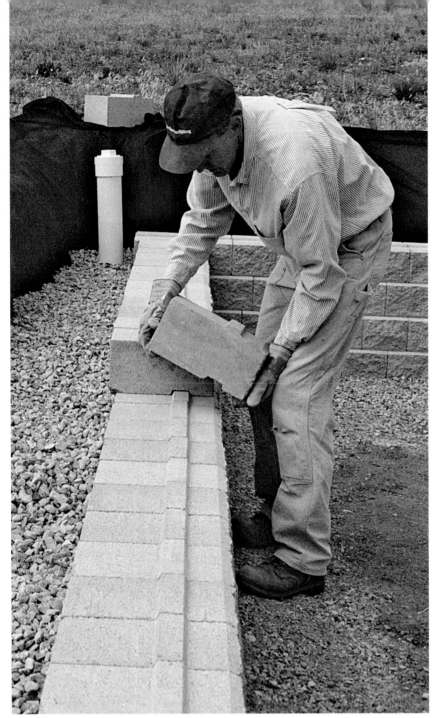

Weatherproof shims handle minor adjustments. For long stretches of wall, the author finds the work goes quickly. But slowdowns are inevitable. When the tops of adjacent blocks deviate by more than ¹⁄₃₂ in., adjustments may be required.

the foot of the wall, I folded the front of the trench fabric back toward the blocks, trimmed the excess, and tamped and sloped soil away from the top of the base course.

Laying Block Is Plain Hard Work

Years ago, I was working on a simple but physically demanding job when a friend said, "We have just what this job needs: strong backs and weak minds." I often thought of his remark as I worked through the simple process of raising the wall. Working outward from a corner, I'd start by sweeping all dirt and debris from the top of an existing row and then begin laying new block in a running bond, a pattern in which each block overlaps the one beneath it by half its width (see the photo at left). Aligning the groove on the bottom of each block with lugs cast into the tops of the blocks on the course below kept each course straight and stepped back evenly. I was often able to work long sections of wall without hitting a snag, but adjustments were made to keep the wall straight and level.

Even though the blocks are consistent in width, it was often necessary to cut blocks to maintain the running bond pattern from course to course. I used my circular saw and a diamond blade to get predictable results (see the sidebar p. 248). Once I completed a row of block, I backfilled and tamped the drainage stone up to the top of the blocks. Wherever the initial excavation had left a sharply defined bank, backfilling was easy: It meant placing crushed stone between the back of the wall and the fabric-covered face of the dig. But in areas where the excavation sloped back too much, I used a temporary plywood divider next to the geotextile to maintain the slope. The plywood kept the size of the drainage channel fairly consistent and let me backfill dirt behind the fabric at the same time I was adding crushed stone in front of it.

As the wall grew higher, I continued to fill and tamp crushed stone in the 12-in. wide channel behind the wall (see the top right photo on p. 249). In addition to channeling water safely away from the back of the wall, the drain system should also minimize efflorescence, a white powder that forms when water leaches minerals out of the concrete. To stabilize and protect

When blocks have to be shimmed, the author uses narrow pieces of sheet aluminum that won't corrode, rot, or wash away over time.

Sources

Anchor Block Co.
8201 Brooklyn Blvd.
Brooklyn Park, MN
55445
800-440-8657
www.anchorblock.com

GeoWestern, Inc.
2701 W. Mansfield Ave.
Englewood, CO 80110
303-761-0883
www.geowestern.com

Keystone Retaining Wall Systems
4444 W. 78th St.
Minneapolis, MN 55435
800-747-8971
www.keystonewalls.com

McKinney Concrete Products
2700 North Freeway
Pueblo, CO 81003
800-332-8512
www.concretecolorado.com

Mirafi
365 South Holland Dr.
Pendergrass, GA 30567
706-693-2226
www.mirafi.com

RisiStone Systems
8500 Leslie St.
Ste. 390
Thornhill, ON L3T 7M8
Canada
800-626-9255
www.risistone.com

Sashco
10300 E. 107th Place
Brighton, CO 80601
800-289-7290
www.sashcosealants.com

Versa-Lok Retaining Wall Systems
6348 Highway 36 Blvd.
Ste. 1
Oakdale, MN 55128
800-770-4525
www.versa-lok.com

Mortarless Design Means Flexibility

As I got close to finishing individual sections of the wall, I found that being able to dry-stack blocks was a real blessing. Instead of pouring a concrete wall and then wishing it were higher or lower, I tried different arrangements and then assessed the wall's appearance. Once my wife and I chose the final heights for different sections, it was easy to make the top row step up or down. I simply ended a row at any point with a corner block so that the rough face showed on two sides.

With the blocks in place, I leveled crushed stone one row below the top of the wall. I then folded the geotextile forward over the stone and up the back of the top row of block (see the drawing on p. 244). With the excess trimmed off, the exposed fabric was covered with soil. Although I could just have sloped the grade to the top of the wall, I formed a healthy swale behind the wall and planted it with native grasses to help control water runoff.

The easiest part of the project was installing the cap. The standard Pisa Revers-A-Cap® is a trapezoid-shaped piece that can be laid in an alternating pattern to form a straight wall or arranged in the opposite way to form a curve. After placing and cutting the caps to fit a section of wall, I aligned their front edges with a string and then glued them in place with a flexible adhesive (Sashco Set 'N Stone). Four quarter-size dabs of adhesive are all it took to set each piece. Finally, I used cast-stone pier caps in a contrasting color to accentuate each corner in the wall.

*Growing up amid the never-ending remodel of a 19th-century house, **Andy Beasley** developed a love of tools and fine craftsmanship. He served as an Air Force pilot for 20 years, mostly flying as an instructor/examiner in supersonic jets. No longer hurtling through the sky at 600 mph—with a mere 3 ft. separating the planes in his formation—Andy has taken on another challenge. He and his wife are finishing their home in rural Colorado.*

Floor framing is cantilevered 6 in. over foundation wall.

Veneer stone lines up with wall sheathing.

Dress Up a Block Wall with a Rock Wall

IF YOU OVERHANG THE FLOOR FRAMING, STONE VENEER CAN LOOK STRUCTURAL
Stone veneers are different from stone foundations in that they don't support the house. But they're still heavy, so they require a footing. Rather than burying your stonework, though, build up the existing footing with concrete block to grade.

8-in. concrete block with cores filled

Wall ties

Mortar fills around and behind the stones.

Blocks are set on footing almost to grade.

■ BY CODY MACFIE

In the old days, foundations of rock or brick were the norm. They looked good and were fairly easy to build. Nowadays, concrete block or poured concrete is the foundation method of choice because they're much faster to build. This newfound speed, however, comes at an aesthetic cost: Concrete is ugly. But you can make a plain-looking block wall into a great-looking rock wall by veneering it with fieldstone. The tools and materials needed are few, and the payoff is huge.

The techniques for veneering are the same for block, poured concrete, or even a wood-frame wall, as are the requirements. Make sure you have sufficient support below the stone (a solid footing), and attach the veneer to the wall with wall ties. If the veneer is a retrofit, you may need to pour an additional footing, usually about 6 in. wide. And for wood-frame walls, you need to add a moisture barrier, such as peel-and-stick roofing membrane or #30 felt paper, to the wood. Wall ties are easy to install if you're laying up a new block wall. For concrete walls or existing block walls, the ties can be attached with a powder-actuated nail gun or with masonry screws.

Although veneering an entire house is best left to a professional, a short foundation veneer, such as the one featured here, is certainly bite-size enough for a nonmason to attempt.

Tight-Fitting, Yet Unmanipulated

There are as many varieties of stonework as there are stonemasons, but most can be lumped into a few patterns (see the sidebar below). Much of my work is in a style called dry stack, which resembles a traditional no-mortar rock wall. When veneered in the dry-stack style, mortar is packed behind the stones as well as in a thin layer around the stones, but the mortar is not visible. Although dry-stack veneer looks rough and

Rubble Patterns

Rubble patterns refer to stonework that appears unmanipulated, rather than cut or chiseled. Whether the mortar is visible or not, there are a few common patterns for laying up stone.

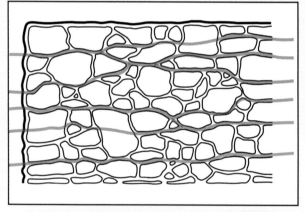

Random rubble has no visible continuous course or bed lines. The stones may fit together tightly but randomly, as featured in the drawing.

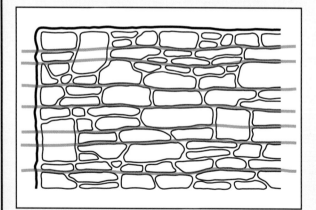

Coursed rubble has a somewhat level bed line with every course. The stones are of varying sizes, but each large stone defines a level bed line.

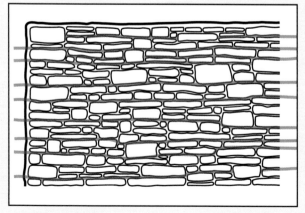

Squared rubble has a level bed line every third or fourth course.

tumble, it's rather precise. The stones fit together tightly, yet look unmanipulated. With jointed-style stonework, you don't have to be as particular because the visible mortar around the stones absorbs the bumps and irregularities.

Good-looking dry-stack veneer is all about tight joints that look natural. You can close gaps between stones by chipping away bumps, by using plugs, or by manipulating the shape of the stone with a hammer and a blunt chisel. Large gaps not only look unnatural but also can allow stones to shift, which creates a weak spot in the wall.

The Most Important Tool Is Space

Being able to look at all the stones to choose the best size, shape, or face for each particular spot—especially the corners—is critical. Because stonework is a mixture of art and grunt labor, plenty of space allows you to take inventory and set aside key stones, such as corners and caps, so that you won't have to switch gears as often. Stopping the process of laying up stone to haul another load can be frustrating.

Start by dumping the stone into a large space near the work area, and shuttle small piles to the wall in a wheelbarrow. The other tools you'll need are a square shovel, a pointing trowel, a mason's trowel, a 4-ft. level, a brick hammer, a 4-lb. rock hammer, a blunt chisel, a plumb bob, a tape measure, and a garden sprayer. If the job is large, rent a cement mixer.

Before You Start, Look Up and Down

Although stone veneer doesn't support weight, a footing must support it. Because the footing is below final grade, stacking 6-in. or 8-in. concrete blocks to just below the final grade line and laying stones on the blocks makes sense. Blocks need to be secure

to make a solid seat for the first course of stones, so set them in mortar. If dirt has covered the footing, dig it out until the footing is exposed.

Verify that the wall to which you are veneering is plumb. A plumb wall speeds the veneering process because you can simply measure the same distance from the wall to the face of the stone as you lay them. If the wall is not plumb, then use a level to make sure the stone faces are plumb.

Mortar Holds the Stone Together

I measure mortar in batches, or the amount that my mixer can mix, that my wheelbarrow can hold, and that I can maneuver around the site. A full batch fills my mixer. I mix either a full batch or a half-batch depending on the weather, my crew size, and proximity to quittin' time. Regardless of whether you use a mixer or a wheelbarrow, the recipe is the same: a 3-to-1 mixture of sand to portland cement. A full batch in my mixer is a half-bag of portland cement and 14 shovelfuls of sand. If you mix in a wheelbarrow, small batches make the mixing much easier. Whether mixer or wheelbarrow, mix the dry ingredients well before adding the water.

Dry-stack mortar can be mixed a bit wetter than jointed style; it should be slightly sticky. To test, take a handful, form it into a loose ball, and throw it into the air. If it stays in a ball, you're ready to go. If it crumbles and doesn't stay in a ball, slowly add water. Be conservative. There is probably more water in the mix than you realize, and if the mix becomes too soupy, you'll need to add more sand and cement. On hot days, mix the mortar a little wet because it tends to dry quickly, especially when sitting in the sun.

A fast-drying variation of this 3-to-1 recipe is to change the cement mix from 100-percent portland to half portland and half type S. Type-S cement is stickier and

TIP

I never build more than 4 ft. or 5 ft. high in one day without using fast-setting (type-S) mortar. Portland-cement-based mortar won't cure enough to hold the weight.

Mortar Binds the Wall

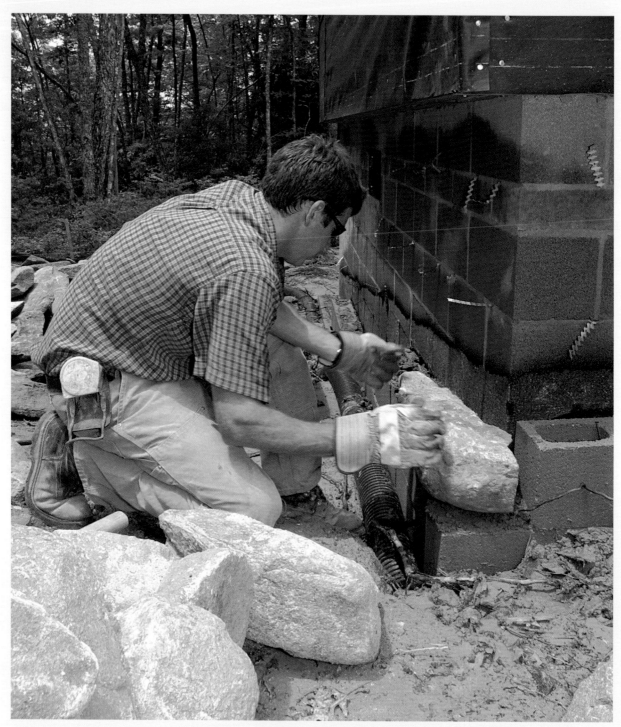

Lay stone on a bed of mortar and pack more behind. The unseen mortar below, beside, and behind the stone holds the wall together, but visible mortar in front will be scratched out.

Trim rock with a blunt hammer. Break bumps off the back and bottom to improve fit and to reduce the chances that the stone will move as rocks are piled on top of each other.

Use shims to set the stones more securely. The rock-fragment shims can be used in one of two ways: either as temporary shims until the mortar sets up or as permanent plugs to fill gaps between stones, giving a tighter appearance.

Pack mortar behind the stones. Mortar holds the stone in place, and wall ties embedded into the mortar tie the stone veneer to the block wall. Lay a mortar bed on top of each stone course for the next course.

sets up faster. I use this recipe when I need to be able to build a wall higher than 4 ft. or 5 ft. in a single day. The stickier mortar adheres well to the stone, and it dries within a couple of hours.

The Craft of Stonework: Cutting and Shaping

The difference between a good-looking wall and a monster has a lot to do with your ability to manipulate a rock. The tighter the stones fit together, the neater the overall wall will look. Good masons know how and where to hit a stone, then where to place it.

To trim the edges of large stones, use a blunt chisel; keep the brick hammer sharp for trimming the edges of smaller stones. If you are unhappy with how the stone looks on the wall, take it down and trim it the way you want it, or simply find another stone.

For a rustic look, minimize surface chiseling; don't trim the textured faces you want exposed. On this job, the homeowners wanted an organic, native stone with a lot of texture, natural weathered color, rigid lines, and shadowed indentations, so I left the faces alone. The sides and tops of the rocks, however, aren't exposed, so I was fairly liberal in trimming around the edges.

Without surface chiseling, the face of the wall will vary somewhat. I set the face of each stone roughly 6 in. from the block wall. Some surface lumps or dimples will be closer or farther. The main body of the stone aligns, and the surface irregularities provide texture.

4-ft. level

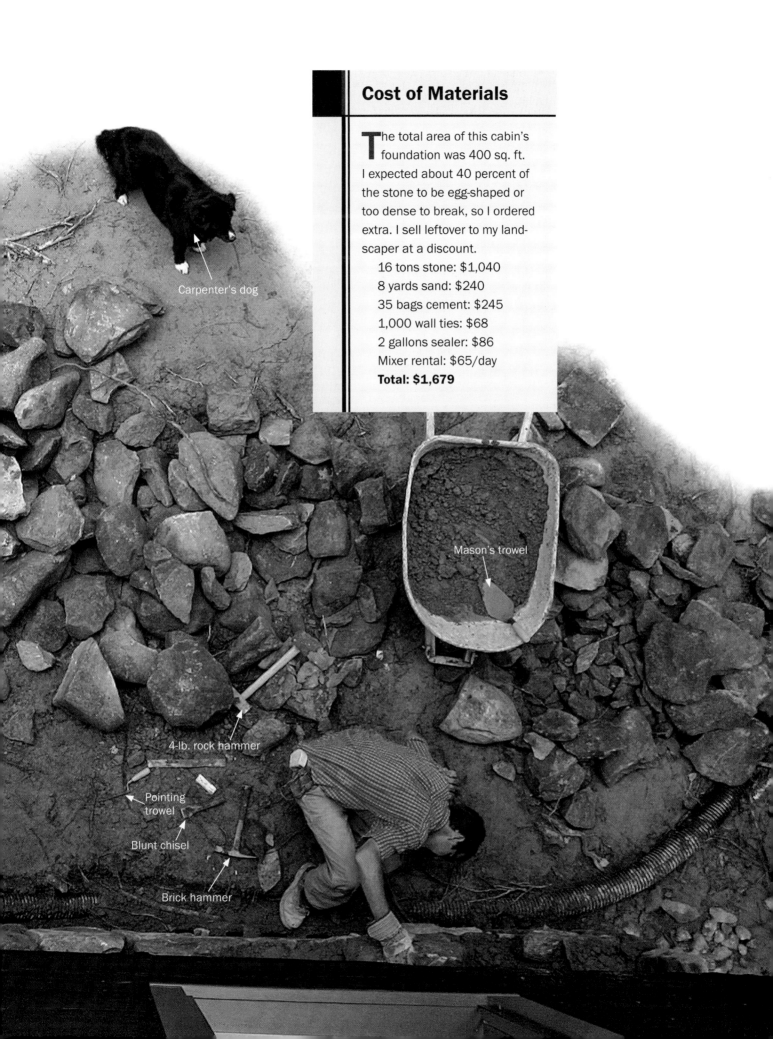

Cost of Materials

The total area of this cabin's foundation was 400 sq. ft. I expected about 40 percent of the stone to be egg-shaped or too dense to break, so I ordered extra. I sell leftover to my land-scaper at a discount.

16 tons stone: $1,040
8 yards sand: $240
35 bags cement: $245
1,000 wall ties: $68
2 gallons sealer: $86
Mixer rental: $65/day
Total: $1,679

Carpenter's dog

Mason's trowel

4-lb. rock hammer

Pointing trowel

Blunt chisel

Brick hammer

Because cornerstones have two exposed faces, it's a good idea to choose them first. And because the corners dictate the course lines, that's where I start. After setting a couple of alternating cornerstones, I lay a long base of horizontal stones before building up. I never build more than 4 ft. or 5 ft. high in one day without using fast-setting (type-S) mortar. Portland-cement-based mortar won't cure enough to hold the weight. When placing each stone, orient it so that the thickest part is on the bottom, which keeps it from kicking out when weight is stacked on top. Make sure the stone doesn't shift before you fill in with cement. And don't trim rocks while they are resting on the wall; trimming can loosen surrounding stones before they are set.

Long Stones Make the Wall Look Stronger

Stones often are packaged in similar shapes: long horizontal stones, nuggets, rounded fieldstones, and so on. A pattern that I like is a mixture of 20 percent to 40 percent fieldstones and 60 percent to 80 percent horizontal stones, but the final pattern is somewhat dependent on how the stone yard packages the stone. For this job, I bought the stone for the project in bulk to get a more random selection of rock shapes, because the homeowners didn't want the wall to have a formal pattern. Even for a random pattern, though, I follow a couple of rules.

Rule #1: Always bridge vertical joints with the stones in the next course. Running vertical joints are not pleasing to the eye and eventually can crack if the foundation settles or shifts.

Rule #2: Alternate corners to the left and right as you set each course. Even with a rustic pattern such as this one, structure demands that the quoins, or large cornerstones, alternate. Although I didn't pull strings from the wall ends for a straight corner, I did take care to choose cornerstones with faces at right angles to one another.

Cap the Wall

Because this veneering project tucks under cantilevered framing, a perfect cap isn't as critical. However, if a veneer projects beyond the siding, a flat cap with the same type of stone gives the wall a finished look and allows it to shed water.

If you know the veneer will need a finished cap, make sure you leave enough room for it. Up to 2 in. more than the thickness of the capstones is enough space to angle the stone away from the house to shed water. Tap capstones with a rubber mallet to set them in position. Make sure the capstones are level. One easy way is to snap a chalkline across the wall before you set the last course of stone.

Finish with a Brush and a Sealer

As you lay the stones, packing mortar behind them to set each one and to hold the wall ties, some mortar will make its way to the surface cracks. After a couple of hours of curing, scratch away this excess with a small pointing trowel. The mortar should crumble and fall out. Scratching too soon may smear cement on the edges of the stones or compromise the integral structure of the hidden mortar bed. With dry-stack veneering, you don't need to finish the joints, so after scratching out excess mortar, brush the joints with a small broom.

After a few days, the mortar should be cured fully and ready for a waterproofing sealer. I like Sure Klean® Weather Seal Siloxane PD (www.prosoco.com; 800-255-4255). Waterproofing keeps moisture out of the basement and also prevents efflorescence. Apply sealant to the stone with a garden sprayer. The most important place to seal is the top of the wall (the cap), because this spot gets the most water.

Cody Macfie is a second-generation stonemason as well as a freelance writer. He lives and works in the mountains of western North Carolina.

Alternate the Cornerstones

Because cornerstones have two faces exposed, pick them carefully. To make a corner strong, the stones should alternate directions. Build up the corners, then work sideways into the field.

Rake out the semidry mortar. After a few hours, the mortar is dry enough to remove all that is visible. Use a pointing trowel and go deep. There should be no visible mortar in a dry-stack veneer wall.

Keep the rocks plumb. Regardless of how thick the rocks are, the faces should be in the same plane. If the foundation wall is plumb, you can measure to the face of the rocks consistently.

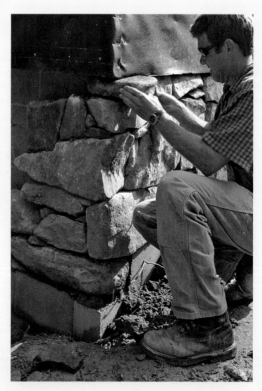

The capstone takes a little planning. Select the capstones before you place the preceding course. Because the siding will hang down an inch or so, there's some wiggle room that can be filled with mortar.

A Concrete-Paver Patio from the Bottom Up

■ BY DICK HENRY

On those first warm days of spring, a patio tucked into a sunny corner of the backyard begs you to spend time outside. Throughout summer and into fall, that versatile outdoor room can be a solitary retreat or a place to entertain guests. Regardless of the season, a flat and firm patio made from simple materials adds value to your home.

Traditionally, patios have been made of brick, flagstone, or poured concrete, but I prefer installing concrete pavers. Pavers are widely available, and you can buy them in different shapes and colors. The cost of a patio made with concrete pavers is roughly twice that of poured concrete and comparable to clay-brick pavers. Site conditions and labor costs vary widely, but here in North Carolina, where I work, the installed cost of a paver patio runs between $10 and $20 per sq. ft. (see the sidebar on p. 264). I like the durability of concrete pavers, and their consistent size makes for ease of installation.

When you're installing a paver patio or walk, it's what you do before the pavers go down that makes all the difference. If you plan carefully (see "Before You Begin" on the facing page) and prepare a good base, you'll set the stage for a patio that looks great and will stay that way for a long time.

Dick Henry started his business in 1977. Five years later, he hired his 81-year-old grandfather for general labor and inspiration. Dick has a bachelor's degree in landscape architecture from North Carolina State University. A fan of big trees, he aspires to visit old-growth forests around the world.

Before You Begin

- Use spray paint to outline the patio on the ground.
- Before excavating, call a local utility or service to locate underground hazards.
- Cover or protect sensitive plants.
- Find a flat, convenient location to unload pavers, stone, sand, etc.
- Do you need to remove soil? If so, decide where it will go.
- Patios change the way water drains around your house. Determine what drainage alterations will be necessary to move water away from the house.
- Plan outdoor lighting and power locations, and install underground lines as necessary.

Regardless of the season, a flat and firm patio made from simple materials adds value to your home.

Stringlines Guide the Excavation

To end up with a flat patio that drained water away from the house, I created a virtual patio of string as a reference during construction. I chose 7 in. below the back-porch floor, the height of an average step, as the patio's high point. I tied a string to the first grade pin at this height, and to guard against losing the mark, I wrapped duct tape at the same height. At the patio's far end, I drove another grade pin and stretched a string between the two pins. Measuring down from the string tells whether the soil below is too high, too low, or just right.

¾-in.-dia. by 30-in.-long steel grade pin

Extend the stone base 10 in. to 12 in. beyond the patio. This allows for minor adjustments in the size of the patio and provides stability at the edge.

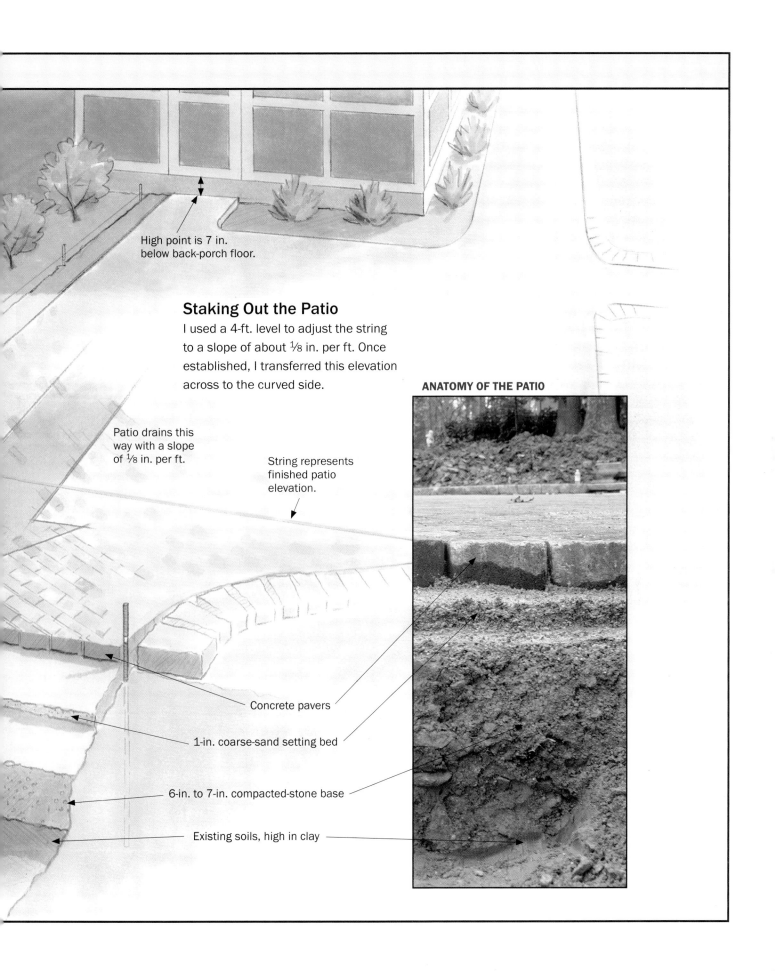

High point is 7 in.
below back-porch floor.

Staking Out the Patio

I used a 4-ft. level to adjust the string
to a slope of about ⅛ in. per ft. Once
established, I transferred this elevation
across to the curved side.

ANATOMY OF THE PATIO

Patio drains this
way with a slope
of ⅛ in. per ft.

String represents
finished patio
elevation.

Concrete pavers

1-in. coarse-sand setting bed

6-in. to 7-in. compacted-stone base

Existing soils, high in clay

The Stone Base Is Critical

For a lasting job, nothing is more important than getting the stone base firm and flat. I excavate the site 9 in. to 10 in. below the finished grade to make room for the stone base. Slope the excavation to promote drainage, following the stringlines. A Bobcat makes quick work of moving stone; a plate compactor packs and flattens the base.

Spread the stone in thin layers. The ABC stone we used for the base includes all gradations of stone from dust to 1½-in. dia. gravel. This mix can be packed flat and dense without losing its drainage properties. Apply the base in several 2-in. to 3-in. layers, wetting it down to aid compaction. Repeat the layering process until the base is 3 in. from the finished elevation.

I Hit a Soft Spot

After excavating to the desired depth, I still have unstable soil to take care of. I install geotextile fabric over this area. The geotextile helps to spread the weight of the ensuing stone base, creating a stable surface.

Use Coarse Sand for Bedding the Pavers

A maximum of 1 in. of clean, coarse sand is installed on top of the stone base. Initially, I bring in a little more sand than I need. After running the plate compactor, I'll end up close to my target elevation, but I don't worry about perfection at this stage. Perfection comes with the final screeding.

Fine-tune the setting-bed layer. After spreading and compacting the coarse sand over the base, I extend stringlines from the grade pins to identify the finished height of the patio. Then we use a paver and a 4-ft. level (or other straightedge) to fine-tune the setting bed. The final smoothing fluffs the sand a little after compaction so that the pavers bed nicely in the sand.

Circular Pattern Starts with a Kit

We chose a circular pattern for this patio because it complements the curve in the retaining wall. These pavers came from Belgard (Oldcastle Architectural Product Group; 800-899-8455; www.belgard.biz), so we bought their starter kit for circular pavers. The kit includes a couple of sizes of wedge-shaped pieces, which can be used in various combinations to create the first dozen or so rings. After that, the square and rectangular pavers continue the pattern. Like most paver manufacturers, Belgard provides a good installation guide for anyone unfamiliar with the process.

Start in the Center

Where's the center? Matching the paver pattern to the retaining wall would have placed the pattern's center too close to the house, so we started closer to the center of the patio. When installing pavers in running rows, we're careful not to line up the joints.

Make Straight Cuts in Place

Straight runs are easy. Snap a chalkline, then use a cutoff saw with a diamond blade to make a quick scoring cut along the top of the paver. Make a final pass, cutting through the pavers and slightly into the sand.

Cutting Curves Is Trickier

Mark curves with ¾-in. braided rope, which bends and holds a curve. When the curve is formed, mark it with a pencil and score it lightly with the saw to make a smooth curve. Take each piece out and complete the cut.

Lock the Pavers into Place

Sweep sand into the cracks. We spread clean, coarse sand over the patio and sweep it into the cracks. Then we vibrate the pavers with a compactor to lock the pavers into place. We repeat the process at least one more time: spread, sweep, and vibrate.

Anchor Pavers around the Edges

On this job, we used a masonry border (from Belgard) with plastic inserts that poke into the pavers and then are spiked into the ground. My favorite edge restraints are the 8-ft. polyethylene strips, which can form curves if you snip the intermediate tabs.

CREDITS

The articles compiled in this book appeared in the following issues of *Fine Homebuilding*:

Table of Contents: Photo on p. vi (left) by Charles Bickford, courtesy of *Fine Homebuilding*, © The Taunton Press, Inc.; p. vi (right) by Christopher Ermides, courtesy of *Fine Homebuilding*, © The Taunton Press, Inc.; p. vii (left) by Steve Culpepper, courtesy of *Fine Homebuilding*, © The Taunton Press, Inc.; p. vii (right) by Roe A. Osborn, courtesy of *Fine Homebuilding*, © The Taunton Press, Inc; p. viii (left) courtesy of Koch Materials Company; p. viii (right) © Harold Shapiro; p.1 (left) by Roe A. Osborn, courtesy of *Fine Homebuilding*, © The Taunton Press, Inc.; p. 1 (right) by Daniel S. Morrison, courtesy of *Fine Homebuilding*, © The Taunton Press, Inc.

p. 4: Avoiding Common Masonry Mistakes by Thor Matteson, issue 103. Photos on 5, 8, 15 © Thor Matteson. Drawings © Mark Hannon.

p. 16: Working with Rebar by Howard Stein, issue 137. Photos on p. 17 and 21 by Charles Bickford, courtesy of *Fine Homebuilding*, © The Taunton Press, Inc.; p. 22 © Charles Bickford; p. 19 and 20 © Howard Stein. Drawings © Christopher Clapp.

p. 25: Placing a Concrete Driveway by Rocky R. Geans, issue 102. All photos by Roe A. Osborn, courtesy of *Fine Homebuilding*, © The Taunton Press, Inc. Drawing on 26 © Vince Babak.

p. 36: Drilling Concrete? Get a Rotary Hammer by David Crosby, issue 162. All photos © Scott Phillips except p. 37 by Chris Green, courtesy of *Fine Homebuilding*, © The Taunton Press, Inc.; fasteners, pp. 38– 41, courtesy of Concrete Fastener Systems Inc.; p. 41 (far left) By Scott Phillips, courtesy of *Fine Homebuilding*, © The Taunton Press, Inc.; bits, p. 42, courtesy of Bosch Power Tools. Drawing © Bruce Morser.

p. 44: A Solid Deck Begins with Concrete Piers by Rick Arnold, issue 180. All Photos by Christopher Ermides, courtesy of *Fine Homebuilding*, © The Taunton Press, Inc. except p. 49 (bottom right) courtesy of Ardisam Inc. Drawings by Dan Thornton, courtesy of *Fine Homebuilding*, © The Taunton Press, Inc.

p. 54: Soil: The Other Half of Foundation by Robert M. Felton, issue 136. Photo on 55 by Tom O'Brien, courtesy of *Fine Homebuilding*, © The Taunton Press, Inc; p. 58 courtesy of S&ME, Raleigh, NC; p. 60 courtesy of Avongard. Drawings © Christopher Clapp.

p. 62: Start with Batterboards by Jim Blodgett, issue 169. All photos © Hugh Lentz. Drawing by Dan Thornton , courtesy of *Fine Homebuilding*, © The Taunton Press, Inc..

p. 66: Pouring Concrete Slabs by Carl Hagstrom, issue 83. All photos by Rick Zeigner, courtesy of *Fine Homebuilding*, © The Taunton Press, Inc.

p. 76: Building a Block Foundation by Dick Kreh, issue 15. All photos © Dick Kreh.

p. 86: Laying up Concrete Block by John Carroll, issue 111. All photos by Steve Culpepper, courtesy of *Fine Homebuilding*, © The Taunton Press, Inc., except photo on p. 88 (bottom) by Scott Phillips, courtesy of *Fine Homebuilding*, © The Taunton Press, Inc.

p. 96: Forming and Pouring Footings by Rick Arnold and Mike Guertin, issue 119. All photos by Roe A. Osborn, courtesy of *Fine Homebuilding*, © The Taunton Press, Inc.

p. 104: Forming and Pouring Foundations by Rick Arnold and Mike Guertin, issue 120. All photos by Roe A. Osborn, courtesy of *Fine Homebuilding*, © The Taunton Press, Inc.

p. 116: Footing and Foundation in One Pour by Jim Blodgett, issue 165. All photos © Hugh Lentz. Drawings by Dan Thornton, courtesy of *Fine Homebuilding*, © The Taunton Press, Inc.

p. 122: Insulated Concrete Forms by Andy Engel, issue 128. All photos by Andy Engel, courtesy of *Fine Homebuilding*, © The Taunton Press, Inc.

p. 138: A Fast Foundation Addition by Rick Arnold, issue 170. All photos by Roe A. Osborn, courtesy of *Fine Homebuilding*, © The Taunton Press, Inc. except p. 141 (center), 143, and 144 (left) by Justin Fink, courtesy of *Fine Homebuilding*, © The Taunton Press, Inc. All drawings by Dan Thornton except p. 139 by Brian Pontolilo, courtesy of *Fine Homebuilding*, © The Taunton Press, Inc.

p. 146: Building a House of Insulated Concrete Block by Fred Leadbeater, issue 132. All photos © Fred Leadbeater, except p. 147, 154-55 by Scott Gibson, courtesy of *Fine Homebuilding*, © The Taunton Press, Inc. Drawings © Christopher Clapp.

p. 156: The Stay-Dry, No Mold Finished Basement by Andy Engel, issue 169. All photos by Charles Bickford, courtesy of *Fine Homebuilding*, © The Taunton Press, Inc. Drawings © Toby Welles @ Design Core.

p. 164: Moisture-Proofing New Basements by Bruce Greenlaw, issue 95. Photos on 164 courtesy of Koch Materials Company; p. 167, 172, and 174 (left) by Bruce Greenlaw, courtesy of *Fine Homebuilding*, © The Taunton Press, Inc.; p. 171 courtesy of W.R. Grace & Company; p. 173 © Brent Anderson; p. 174 (right) courtesy of Dow Chemical Company.

p. 176: Sealing a Crawlspace by Larry Janesky, issue 153. All photos © Harold Shapiro except fastener photo p.181 by Roe A. Osborn, courtesy of *Fine Homebuilding*, © The Taunton Press, Inc. Drawings © Mark Hannon.

p. 184: Details for a Dry Foundation by William B. Rose, issue 111. Illustrations by Christopher Clapp, © The Taunton Press, Inc.

p. 194: Keeping a Basement Dry by Larry Janesky, issue 134. Photos on p. 195, 196, and 201 by Tom O'Brien, courtesy of *Fine Homebuilding*, © The Taunton Press, Inc.; pp. 197 and 199 © Larry Janesky; pp. 202 and 203 © Harold Shapiro. Illustrations by Rick Daskam, © The Taunton Press, Inc.

p. 204: Foundation Drainage by David Benaroya Helfant, issue 50. Photos by Charles Miller, courtesy of *Fine Homebuilding*, © The Taunton Press, Inc. Illustrations © Frances Ashforth.

p. 214: When Block Foundations Go Bad by Donald V. Cohen, issue 117. Photos © David V. Cohen. Illustrations by Christopher Clapp, © The Taunton Press, Inc.

p. 223: Retrofitting a Foundation by William Anthony, issue 129. Photos on pp. 224-231 (left and top) by Roe A. Osborn, courtesy of *Fine Homebuilding*, © The Taunton Press, Inc.; p. 231 (bottom) © Jack Venning.

p. 232: Four Retaining Wall Choices by Eric Nelson, issue 173. Photos by Brian Pontolilo, courtesy of *Fine Homebuilding*, © The Taunton Press, Inc. except p. 233 and 235 (bottom) courtesy of Walton & Sons Masonry Inc., Mountain View, Calif.; p. 235 (top) by Gary M. Katz, courtesy of *Fine Homebuilding*, © The Taunton Press, Inc.; p. 236 courtesy Versa-Lok; p. 237 courtesy Allan Block; p. 241 courtesy Ted Lane Design/Build, Des Moines, Iowa. Drawings © Don Mannes.

p. 242: Building a Modular Retaining Wall by Andy Beasley, issue 141. All photos © Andy Beasley. Drawings © Gary Williamson.

p. 252: Dress Up a Block Wall with a Rock Wall by Cody MacFie, issue 177. All photos by Daniel S. Morrison, courtesy of *Fine Homebuilding*, © The Taunton Press, Inc.

p. 262: A Concrete-Paver Patio from the Bottom Up by Dick Henry, issue 170. Photos by Chris Green, courtesy of *Fine Homebuilding*, © The Taunton Press, Inc. except p. 263 © Tommy McNabb. Drawings by Dan Thornton, courtesy of *Fine Homebuilding*, © The Taunton Press, Inc.

INDEX

Taunton's FOR PROS BY PROS Series
A Collection of the best articles from Fine Homebuilding magazine

OTHER BOOKS IN THE SERIES:

Taunton's For Pros By Pros:
RENOVATING A BATHROOM
ISBN 1-56158-584-X
Product # 070702
$17.95 U.S.
$25.95 Canada

Taunton's For Pros By Pros:
BUIILDING ADDITIONS
ISBN 1-56158-699-4
Product # 070779
$17.95 U.S.
$25.95 Canada

Taunton's For Pros By Pros:
BUIILDING STAIRS
ISBN 1-56158-653-6
Product # 070742
$17.95 U.S.
$25.95 Canada

Taunton's For Pros By Pros:
BUILT-INS AND STORAGE
ISBN 1-56158-700-1
Product # 070780
$17.95 U.S.
$25.95 Canada

Taunton's For Pros By Pros:
EXTERIOR SIDING,
TRIM & FINISHES
ISBN 1-56158-652-8
Product # 070741
$17.95 U.S.
$25.95 Canada

Taunton's For Pros By Pros:
FINISH CARPENTRY
ISBN 1-56158-536-X
Product # 070633
$17.95 U.S.
$25.95 Canada

Taunton's For Pros By Pros:
FOUNDATIONS AND
CONCRETE WORK
ISBN 1-56158-537-8
Product # 070635
$17.95 U.S.
$25.95 Canada

Taunton's For Pros By Pros:
RENOVATING A KITCHEN
ISBN 1-56158-540-8
Product # 070637
$17.95 U.S.
$25.95 Canada

Taunton's For Pros By Pros:
FRAMING ROOFS
ISBN 1-56158-538-6
Product # 070634
$17.95 U.S.
$25.95 Canada

Taunton's For Pros By Pros:
BUILDING PORCHES
AND DECKS
ISBN 1-56158-539-4
Product # 070636
$17.95 U.S.
$25.95 Canada

Taunton's For Pros By Pros:
ATTICS, DORMERS,
AND SKYLIGHTS
ISBN 1-56158-779-6
Product # 070834
$17.95 U.S.
$25.95 Canada

Taunton's For Pros By Pros:
FRAMING FLOORS,
WALLS AND CEILINGS
ISBN 1-56158-758-3
Product # 070821
$17.95 U.S.
$25.95 Canada

Taunton's For Pros By Pros:
TILING
ISBN 1-56158-788-5
Product # 070843
$17.95 U.S.
$25.95 Canada

Taunton's For Pros By Pros:
ROOFING, FLASHING
AND WATERPROOFING
ISBN 1-56158-778-5
Product # 070833
$17.95 U.S.
$25.95 Canada

Taunton's For Pros By Pros:
CABINETS AND
COUNTERTOPS
ISBN 1-56158-806-7
Product # 070858
$17.95 U.S.
$25.95 Canada

Taunton's For Pros By Pros:
WINDOWS AND DOORS
ISBN 1-56158-808-3
Product # 070860
$17.95 U.S.
$25.95 Canada

For more information contact our website at: www.taunton.com